Man
Discovers
the Galaxies

Man Discovers the Galaxies

Richard Berendzen
AMERICAN UNIVERSITY

Richard Hart
NATIONAL ACADEMY OF SCIENCES

Daniel Seeley
HARVARD UNIVERSITY

SCIENCE HISTORY PUBLICATIONS

a division of Neale Watson Academic Publications, Inc.

New York   1976

Published by Science History Publications
a division of Neale Watson Academic Publications, Inc.
156 Fifth Avenue, New York 10010

Sole world distributor, excluding the United States, its posses-
sions, and Canada: McGraw-Hill International Book Company

This work was prepared with the support of National Science
Foundation grant No. GY 6685. Any opinions, findings, conclusions or recommendations in said work are those of the
authors and do not necessarily reflect the views of the National
Science Foundation. Except for the rights to material reserved
by others, the Publisher and the copyright owner hereby grant
permission to domestic persons of the United States and Canada
for use of this work without charge in the English language in
the United States and Canada after January 1, 1982. For
conditions of use and permission to use materials contained
herein for foreign publication or publications in other than the
English language, apply to the copyright owner.

First edition 1976

Designer: Ernst Reichl. Manufactured in the U.S.A.

Library of Congress Cataloging in Publication Data

Berendzen, Richard.
    Man discovers the galaxies.

    Bibliography: p.
    Includes index.
        1. Astronomy—History—United States. 2. Astronomy—History—Europe.
    I. Hart, Richard Cullen, joint author. II. Seeley, Daniel, joint author.
    III. Title.
    QB32.B47   520'.9'04   75-4688
    ISBN 0-88202-023-4

Contents

Abbreviations

in References

Ann. Harvard Coll. Obs.	Annals Harvard College Observatory
Ann. N. Y. Acad. Sci.	Annals of the New York Academy of Sciences
Ann. Soc. Sci. Brux.	Annals de la Societe Scientifique de Bruxelles
Ark. Math., Astron, Fys.	Arkiv foer Mathematik, Astronomisch Fyskik
Astron, Gesell.	Astronomische Gesellschaft
Astron. Nachr.	Astronomische Nachrichten
Astrophys. J.	Astrophysical Journal
Bull. Astron. Inst. Neth.	Bulletin of the Astronomical Institutes of the Netherlands
Bull. Astron. Soc. Neth.	Bulletin of the Astronomical Society of the Netherlands
Bull. Nat. Acad. Sci.	Bulletin of the National Academy of Sciences
Bull. Nat. Res. Coun.	Bulletin of the National Research Council
C.R. Acad. Sci.	Comptes Rendus Hebdomadaires des Seances de l'Academie des Sciences
Hale Collection	George Ellery Hale Papers, California Institute of Technology, 1968
Handb. Astrophys.	Handbuch der Astrophysik
Handb. Phys.	Handbuch der Physik
Harvard Bull.	Harvard Bulletin
Int. Astron. Un. Trans.	International Astronomical Union Transactions
J. Brit. Astron. Assoc.	Journal of the British Astronomical Association
J. Hist. Astron.	Journal for the History of Astronomy
Lick Obs. Bull.	Lick Observatory Bulletin
Lowell Obs. Bull.	Lowell Observatory Bulletin
Med. Astron. Obs. Upsala	Meddelanden Fran Astronomiska Observatoire Upsala
Mon. Not. R. Astron. Soc.	Monthly Notices of the Royal Astronomical Society
Mt. Wilson Cont.	Mt. Wilson Contributions
Nord. Astron. Tides.	Nordisk Astronomisk Tideskrift
Proc. Int. Conf. Ed. Hist. Astron.	Proceeding of the International Conference on Education and the History of Astronomy
Proc. Natl. Acad. Sci. USA	Proceedings of the National Academy of Sciences of the United States of America
Proc. R. Soc. Lond.	Proceedings of the Royal Society, London
Publ. Astron. Soc. Pac.	Publications of the Astronomical Society of the Pacific
Publ. Astrophys. Insti. Königstuhl-Heidelberg	Publikationen des Astrophysikalischen Institutes Königstuhl-Heidelberg
Publ. Dominion Astrophys. Obs.	Publications of the Dominion Astrophysical Observatory
Publ. Lick Obs.	Publications of the Lick Observatory
Publ. Yerkes Obs.	Publications of the Yerkes Observatory
Q. J. R. Astron. Soc.	Quarterly Journal of the Royal Astronomical Society
Sci. Am.	Scientific American
Sci. Mon.	Scientific Monthly
Sol. Phys.	Solar Physics
Ups. Astron. Obs. Med.	Upsala Astronomiska Observatorium Meddelanden

Prologue

to the Prologues

Drs. Berendzen, Hart, and Seeley have done a real service by having written a readable and scholarly book dealing mostly with the discovery of the arrangement of our Home Galaxy, the Milky Way System, and with the discovery of the apparently quite basic properties of the Universe of Galaxies in which we find ourselves. The bulk of the text deals with the history-making discoveries by the band of eight or ten great astronomers who led the way—and some who led us astray for awhile—between about 1915 and 1940. This was the period during which Shapley gave the world the general outline of our Galaxy, rounded out by subsequent discoveries of Oort, Lindblad, and Trumpler. It was also the period during which Hubble and others established the true nature of the "spiral nebulae" and other "nebulae" now known to be galaxies in their own right, many not unlike our Home Galaxy. The text before us describes also how Hubble and Humason, building upon the early work of V.M. Slipher, established the velocity-distance relation for galaxies, which in turn proved basic to the viability of the theory of the Expanding Universe.

As one who became a budding Milky Way astronomer in the 1920's, who knew all the actors personally, and who has been watching the developments over more than fifty years of professional activity, I took pleasure in reading how our three young historians of astronomy view these pano-ramic developments. I learned much that was new from my reading of the text. The book was obviously written at the right time. Not only has the dust pretty well settled on the fights and arguments of the 1920's, but the authors had the advantage of having known well several of the principal scientists who were involved. On top of that, they have had access to files of relevant correspondence, that I would not have imagined still existed. It is good to have the whole story together in one place, presented in a most readable form.

I hope that the reader who first encounters this book will follow the pattern of approach that served me well. I read first the Prologues to Sections One, Two, Three, and Four. Next, I skipped through the pages and selected the juiciest and most controversial paragraphs (and references) for my initial reading. And then I leaned back in my easy chair, settled down with a glass of beer, and read at the rate of 25 pages per night.

I hope that many readers may enjoy this book as I did, that they may learn a lot by their reading, and that they too may wish to thank the authors for having written a pleasant and thoroughly instructive historical volume.

Bart J. Bok
PROFESSOR EMERITUS
University of Arizona

Tucson, Arizona
May 20, 1976

Preface

Of all the great themes in astronomy, one of the most profound and provocative has been the discovery of galaxies and of man's place in the cosmic order. Although this story stretches over millennia, the most revolutionary findings have come during the twentieth century. The efforts of the people who made these discoveries—including their thoughts, their methods, and their instruments—comprise the subject matter of this book. Man's developing understanding of the galaxian universe has been characterized by sudden, unexpected discoveries; by long, tedious work; and, sometimes, by false, misleading results. Although the subject matter of astronomy may be impersonal and inanimate, the process of scientific inquiry is a thoroughly human enterprise, and this book stresses the role of people, with all of their strengths and imperfections.

We hope this book will be of interest and use to any reader with an interest in galactic astronomy. The largest group of such individuals probably will be college-level students taking introductory astronomy courses for non-science majors. The book, nonetheless, should also be of use in courses such as physical science for non-scientists, basic astronomy for scientists, physics for both scientists and non-scientists, and the history of science. Likewise, we hope it would be of value in certain teacher training programs for both pre-service and in-service teachers. And we also hope it will be of interest and value to lay readers.

The approach in this book differs from that of conventional textbooks in several respects: This volume does not presume to be comprehensive, surveying all facets of astronomy; rather, it concentrates on a single, highly significant episode in the development of modern science. In addition, it makes extensive use of original, archival information, much of which has never before been published. This information makes it possible to describe with unusually good documentation the events surrounding certain traumatic scientific discoveries, the influences of personalities on scientific research, and the difference between public and private science. A primary aim of this book is to involve the student to some depth in a single, highly important, and unusually exciting problem as it developed in the history of science, and thereby to convey an understanding of both science and scientists.

An asset of the approach used in this book is its inherent flexibility. By including or omitting technical sections, problems, and so on, the subject matter can be adapted for either science or non-science students at various academic levels; moreover, the chapters can be used over a time span ranging from a single lecture to many weeks. Thus, this book is designed primarily to be used in conjunction with a normal college curriculum, not in place of it.

The preparation of this book began in the late 1960's with the support of a grant from the National Science Foundation. The program was then officially entitled "The Case

Studies Project on the Development ot Modern Astronomy" and the Principal Investigator was Richard Berendzen. The Project ultimately led not only to this book but also to several papers published in scientific and educational journals, lectures at professional society meetings, and an international conference on education in and history of modern astronomy. The Project originated when all three of the authors were at Boston University. The early drafts of the materials comprising this book were developed there and were tested there as well as at a number of other colleges and universities in the United States and in Canada.

Because this book rests fundamentally upon archival and unpublished sources, it has been especially necessary for us to receive the assistance and goodwill of numerous individuals and institutions. It is impossible to acknowledge all of those who have assisted us over the years, but we should like to mention the following institutions that were especially helpful: The National Science Foundation; American Astronomical Society; American Institute of Physics; The American University; Boston University; California Institute of Technology; University of California at Berkeley; University of Groningen; Dominion Astrophysical Observatory; Harvard University; Hale Observatories; Huntington Library; Lick Observatory; University of Leiden; University of Pittsburgh; Lund University; Princeton University; Lowell Observatory; the U.S. National Academy of Sciences; the U.S. Naval Observatory Library; the U.S. Library of Congress; and the numerous literary and photographic sources listed in the back of this book. And we would especially like to thank Owen Gingerich of Harvard University and Michael Hoskin of the University of Cambridge, both of whom read early drafts of this manuscript and provided us with many helpful suggestions. We alone, however, take full responsibility for the final version of the man-script.

ONE

The Scale of the Universe

Contents

of Section One

The divine omnipotence in each Milky Way according to Thomas Wright of Durham. (S. Jaki, *The Milky Way*, Science History Publications: New York, 1972)

Prologue

for Section One

In the Beginning . . .

Well before these mighty words were written, men had pondered questions relating to the universe in which they found themselves. Questions concerning the origin, structure, and evolution of the universe are found in even the earliest of man's written records.

Ancient cosmologists were ingenious in their attempts to explain the observable universe—their theories were often logically consistent and their observations were frequently remarkably precise. Modern cosmologists, armed with the full array of weaponry of technological science, have developed a view of the universe that is more amazing than anything the ancients could have conceived. With the advent of powerful astronomical instruments and the under-

3

standing of the physical principles of motion and electromagnetism, we now have, we hope, a more accurate picture of our universe. What is most remarkable about this new understanding, however, is how really recent it is. The discovery of the structure of our local star system and the discovery of other systems (i.e., galaxies) are events that belong entirely to the twentieth century.

The development of scientific theories and the improvement of observations since the eighteenth century have provided the necessary components for the understanding of the universe that we now proudly possess. As we shall see in this section, however, neither theory nor observation could begin to give a complete explanation until the twentieth century; indeed, as we shall see in the last sections, the problems are not yet fully solved.

In this section we shall follow the development of theories on two topics—our Galaxy and external systems. Our Galaxy, i.e., the Milky Way, was first studied systematically in the eighteenth century by England's great immigrant astronomer—Sir William Herschel. At the end of the nineteenth century, studies using more sophisticated methods were again made by a German, H. Von Seeliger, and by a Dutchman, Jacobus C. Kapteyn. Spiral nebulae, which many thinkers believed to be external stellar systems, could not be studied in detail before large telescopes were built; hence their structure was not discovered until mid-nineteenth century, and systematic photographic studies of them began only at the end of the nineteenth century.

The first two decades of the twentieth century brought tremendous progress in many fields of science, astronomy being no exception. Photography, spectroscopy, bigger telescopes, all contributed to the understanding of diverse astronomical questions. Through the synthesis of several astronomical advances a young American astronomer—Harlow Shapley—deduced that the structure of our Galaxy differed radically from the model of Kapteyn, which previously had been generally accepted by the astronomical community.

The conflicts between Shapley's model and that proposed by Kapteyn led to what Otto Struve[1]—himself one of the twentieth century's foremost astronomers—has termed astronomy's "Great Debate." The debate, which took place in Washington, D.C., in 1920, revealed the strengths and weaknesses of both theories. Shapley defended his theory of a radically larger system for our Galaxy. The evidence he used

4

came from studies of globular clusters and pulsating stars, and he simultaneously attacked the proposition that spirals are stellar systems comparable to our own. Heber D. Curtis defended Kapteyn's system, which was based on statistically analyzed counts of stars of varying brightness, and he advocated the "island universe" theory that spiral nebulae are systems much like our own.

Modern commentators do not always agree about the relative importance of the issues that were being debated around 1920. But everyone agrees on one point—the issues discussed were dramatically important and exciting. In 1965, the chairman of the astronomy department at Swarthmore College, Peter Van De Kamp, wrote that "the galactocentric viewpoint developed by Harlow Shapley in 1917 represents a step forward in astronomical thinking and perception analogous to the introduction of the heliocentric viewpoint by Nicholas Copernicus in 1543."[2]

In contrast, a few years later George Abell, the chairman of the astronomy department of the University of California at Los Angeles, remarked in his famous textbook that "the realization, only a few decades ago, that our galaxy is not unique and central in the universe ranks with the acceptance of the Copernican system as one of the great advances in cosmological thought."[3] Obviously, from either point of view, the debate was concerned with cosmological questions of immense significance.

Although the "great debate" did not resolve the disagreements between the systems, it did clarify the issues and demonstrated that there were two basic areas of disagreement over the structure of the universe. Each of these subjects was a complete study in itself, but, at the time of the debate, they could not be divorced; the internal consistency of each theory demanded that both topics be discussed.

In the "great debate" each participant discussed one of the topics at length, treating the other only briefly. Shapley, who was most concerned about the scale and structure of the Milky Way, emphasized that issue; in contrast, Curtis, intrigued with the nature of spiral nebulae, stressed that topic. Clearly they interpreted the actual subject of their discussion (and of this section)—"The Scale of the Universe"—differently. As it turned out, each debater was partly right and partly wrong in his views. In order to approach the debate, we must first discuss man's initial attempts to find a solid, scientific explanation of the scale and structure of the universe.

5

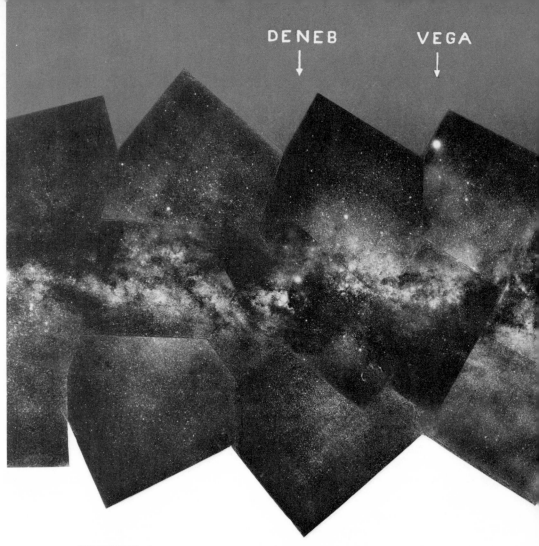

DENEB   VEGA

CHAPTER 1

Early Theories of the Universe

Philosophical Models

Early peoples invariably have seen in the Milky Way the mirror of their own worlds: to Egyptians, it was wheat spread by Isis; to Incas, golden star dust; to Eskimos, a band of snow; to Bushmen, campfire ashes; to Arabs, a river; to Polynesians, a cloud-eating shark; to Teutons, the way to Valhalla; to Iroquois, the path to Ponemah; to Christians, the road to Rome. As we shall see, the ancients were not the only people to impose their personal beliefs upon their theories of the Milky Way and the structure of the universe.

6

ANTARES

The northern Milky Way, from Scorpio to Cassiopeia. Photograph of a mosaic of the prints of the *Atlas of the Northern Milky Way* by F.E. Ross and M.R. Calvert. (Yerkes Observatory photograph)

In 1750, Thomas Wright, an Englishman from Durham, proposed a model for the universe that explained some aspects of the sky. Wright held that the universe was contained in a thin shell between two concentric spheres, the stars being distributed in such a way "as to fill up the whole medium with a kind of regular irregularity of objects."[1] Looking along a tangent to the shell, one would see a multitude of stars; the entire sky in that direction would be filled with faint, distant stars to such an extent that the region would appear as a nebulous glow. If the sun were situated halfway through the thickness of this "shell," the nebulous glow would stretch completely around the sky in the direction of the plane of the tangent. If an observer looked through the thin portion of such a universe, he would see fewer stars; hence, the sky would seem sparsely populated in that direction. By this reasoning, Wright concluded that the appearance of the Milky Way resulted from a spherical distribution of stars.

7

Wright's philosophical discourses were motivated mainly by religion; to him the heavens were a magnificent example of God's handiwork. The spherical structure seemed to Wright the most logical one for God to have constructed. In such a model, God, who was the moral center of the universe, could also be the physical center. Although Wright's model was directed more towards a religious metaphysics than towards a scientific model, he was shrewd enough to attribue the glow of the Milky Way to optical effects.

In the recent past, historians have accepted these thoughts as Wright's last opinion on the subject, and have credited him with tremendous foresight into the cause of the appearance of the Milky Way. In 1966, however, later manuscripts of Wright's were found, edited, and published,[2] which showed a drastic revision of his original concepts of the universe.

The first "letter" of his "Second Thoughts" commenced with the Spanish proverb, *"El subio mucha consigo, el necio no"* (The wise alter their counsel, the foolish do not).[3] Perhaps Wright felt that his revolutionary changes in approach, which we shall soon discuss, required justification.

It is apparent from Wright's manuscript that he had been greatly impressed by the earthquake and the repeated shocks that took place in Lisbon in 1755. He believed they were caused by "a suppos'd abyss of waters or other matter in the center of the Earth . . . inclosed within the terrene crust of mundain shell & which may be well concev'd to be of no great depth or considerable thickness, with regard to ye internal and more expanded submundain space."[4] The earthquake produced waves in the liquid core, reasoned Wright, which bounced back and forth repeatedly but with diminishing force.

Reasoning by analogy with. the spherical crust of the earth, Wright concluded that the celestial sphere must also be solid.

Upon . . . convincing ye idea of a central globe and a circumambient sphere of waters floating around it, I ventured one step farther and was willing to imagine, that in larger orbs, and more immense globes of matter, there might possibly be also vast regions of the air or aether with central spheres of fire and other bodies included . . .

Hear then give me leave to look up again to ye stars; for this idea immediately presented itself to my ready imagination that a new and very possible modification of the visible universe might be derived from it, beleiving it not improbable that the visible heavens or stary firmament might prove to be no other than a solled orb of this stupendous nature and the fixd stars no more than perpetual lumi-

8

nation or vast erruptions & if refulgent or inflammable mater promiscuously distributed as celestial volcanoes all around the starry regions emiting an etherial & intense fire of various magnituds but remov'd for some infinitely wise purpose of the Creator, to so indefinite a distance as to be far with out ye reach of human arts to ascertain but by ye mental eye of reason only.[5]

According to Wright, various celestial phenomena could be explained by such a model. New stars (i.e., novae) might be eruptions of volcanoes, and perhaps comets are glowing material ejected from the volcanoes. He explained variable stars as being caused by the "ebbing and flowing of sidereal flame within the volcanoes." And the appearance of the Milky Way, he claimed, was the result of sidereal fire overflowing like lava. Obviously, he stretched the analogies of the terrestrial model to astronomical limits.

Why Wright returned to the ancient concept of a solid celestial sphere with the Sun at the center instead of eccentrically placed is difficult to tell. His philosophies were always attempts at presenting consistent theories of the moral and physical universes, incorporating coincident centers. Perhaps he felt compelled by the desire for an order of increasingly greater systems, each reproducing the other in form but not in size. Such an array of higher-order systems leads naturally to the concept of infinity in the physical universe, which would be the counterpart of God in the moral universe.

Thomas Wright's illustration of the confinement of stars between parallel surfaces. (S. Jaki, *The Milky Way*, Science History Publications: New York, 1972)

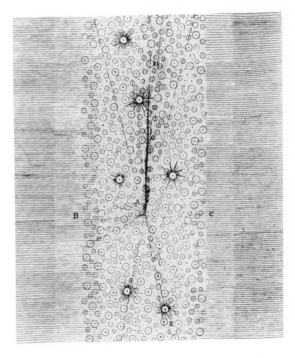

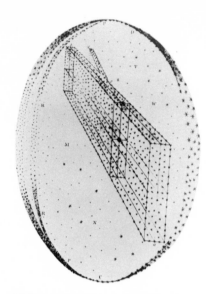

William Herschel's 1784 model of the Milky Way. (S. Jaki, *The Milky Way*, Science History Publications: New York, 1972)

William Herschel's model of the Milky Way based on star-gauges. (M. Hoskin, *William Herschel and the Construction of the Heavens*, Oldbourne: London, 1963)

In 1775, Immanuel Kant elaborated on the Englishman's theory, but largely misinterpreted his model. Wright had believed, at first, that the stars were distributed randomly in the thin spherical region where they were found. The description that Kant had read, however, misrepresented Wright's view and stated that the universe had an order similar to that of the solar system, but of a larger scale and involving many more objects.[6] Hence, the universe envisioned by Kant was a multitude of stars rotating about a common center and nearly all in the same plane. Thus, the distribution of the stars, and the appearance of the heavens, would be similar to that proposed by Wright, but their prescribed motions would be different.

Kant recognized a problem inherent in his theorized movements of the stars: if the stars moved, why do they seem to be fixed, i.e., show no motion in the sky? The answer he provided was reasonable: "It is either only excessively slow, arising from the great distance from the common centre around which the stars revolve, or it is due to mere imperceptibility, owing to their great distance from the place of observation."[7]

Elaborating on Wright's ideas, Kant incorporated into his theory of the universe the small elliptical, luminous patches, which at that time were called nebulous stars. He reasoned that if the Milky Way existed as a disk of stars, might not other flat aggregations of stars also occur? And if these aggregations are as far, relative to their size, from the Milky Way as the individual stars are from each other, then there ought to be small luminous patches, these patches being more or less elliptical, depending on how much they are tilted to our line of sight. Because of his philosophical desire for order, and some observational evidence, Kant convinced himself of the existence of other universes outside the Milky Way: ". . . and if conjectures, with which analogy and observation perfectly agree in supporting each other, have the same value as formal proofs, then the certainty of these systems must be regarded as established."[8]

Scientific Models

Perhaps the first important scientific research on the size and shape of the universe began in the latter part of the eighteenth century when William Herschel attempted to utilize systematic methods, rather than conjecture, to attack the problems of cosmic structure.[9] Herschel's method of fathoming relative distances was simple. He assumed that stars were uniformly distributed throughout the stellar system and that he could see the borders of that system. Then the ratio of the cube roots of the numbers of stars seen in the field of his telescope in given directions reflected the relative distances.

Herschel made extensive counts, which he called "star-gauges," in 683 regions of the sky. From his gauges, he was able to arrive at a rough shape of the universe. His results confirmed Kant's speculation of an elongated universe, although Herschel concluded that the boundary is irregular, with many protrusions.

Herschel was also interested in the nebulous stars mentioned by Kant, and in fact he discovered many new ones himself. Moreover, some objects that previous observers had classified as nebulous were discovered to be clusters of stars when viewed through Herschel's large telescopes. The fact that these nebulous clouds could be resolved into individual stars led Herschel to believe that all nebulosity represented distant stellar systems, which would be resolvable with larger telescopes. This conviction brought him to conclude that all nebulous stars were star systems. In his 1785 paper, Herschel stated:

Portrait of William Herschel. (Yerkes Observatory photograph)

William Herschel's 40-foot focal-length telescope. (Yerkes Observatory photograph)

As we are used to call the appearance of the heavens, where it is surrounded with a bright zone, the Milky Way, it may not be amiss to point out some other very remarkable Nebulae which cannot well be less, but are probably much larger than our own system; and, being also extended, the inhabitants of the planets that attend the stars which compose them must likewise perceive the same phaenomena. For which reason they may also be called milky ways by way of distinction.[10]

He soon discovered, however, that he was unable to explain certain phenomena. In 1790, a crisis arose when he found a planetary nebula symmetrically surrounding a star. Convinced that the nebulosity was associated with the star, Herschel had to conclude that it was not just a stellar system.[11] His previous firm belief that all nebulous clouds are composed of stars was destroyed.

Towards the end of his life, Herschel re-examined his early work and the postulates upon which it had been based. In many cases he changed his mind or at least revised his opinion. He realized that his largest telescope was unable to penetrate to the borders of our Galaxy. At best, the gauges gave only a general idea about the structure of the Milky Way; details of its extent could not be determined that way.

After a lifetime of systematic, careful observing, while making progress and voicing stimulating ideas, Herschel still felt great uncertainty about the nature of nebulae and the size and shape of the universe.

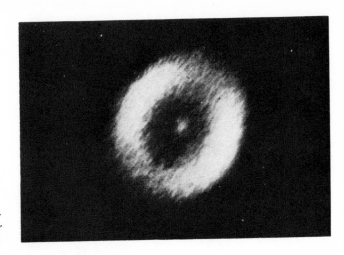

An early photograph (1897) photograph of a planetary nebula. (*Popular Astronomy, 5,* 1897)

After expanding the limits of men's knowledge by both observation and speculation, he reversed himself and again made these limits vague and uncertain.

Herschel's instrumentation was not surpassed for decades. His reflecting telescopes were able to penetrate to distances far beyond the reach of the refractors of his day because the art of lens grinding was then limited to apertures small compared with Herschel's 48-inch mirrors.

In 1845, however, William Parsons, the Earl of Rosse, completed a reflecting telescope with a 72-inch mirror![12] Equipped with an instrument more powerful than Herschel's, Lord Rosse discovered that some nebulae had a spiral structure. He was also able to distinguish individual stars in several nebulae that Herschel had failed to resolve. These additional observations led to a revival of the old theory of external galaxies.

But the fundamental problem still remained—it undercut interpretations of Lord Rosse's observations

The remains of Herschel's 40-foot telescope in the garden of Observatory House, Slough, England. (H.C. King, *History of the Telescope,* Sky Publishing Co.: Cambridge, Mass., 1955)

Lord Rosse's 6-foot telescope. (H.C. King, *History of the Telescope*, Sky Publishing Co.: Cambridge, Mass., 1955)

as it had Herschel's—that no one had a reliable method for determining astronomical distances beyond the nearest stars. Were the spiral nebulae relatively near to us and just swirling clouds of gas, or were they far from us and extremely large? The answer was not obvious.

The problem of uncertainty of distances is evident in the terms "universe" and "Milky Way Galaxy." With no reliable distances, it was not known until this century whether the Milky Way includes all stellar objects (i.e., is the "universe") or if other objects lie outside it, possibly even rivaling it in size. Star clouds and nebulae might be nearby, within the boundaries of our Galaxy; or perhaps Kant's speculations were valid—spiral nebulae might be systems like the Milky Way.

William Parsons, Third Earl of Rosse. (H.C. King, *History of the Telescope*, Sky Publishing Co.: Cambridge, Mass., 1955)

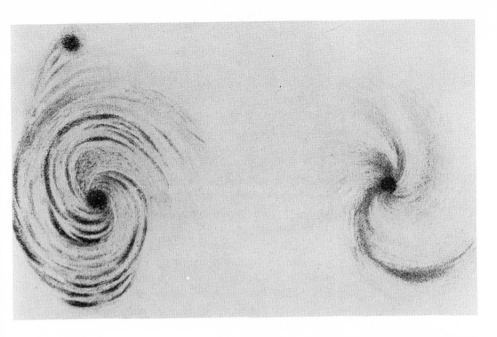

Two sketches of spiral nebulae by Lord Rosse, c. 1850.
(A. Berry, *A Short History of Astronomy*, J. Murry: London, 1898)

CHAPTER 2

The Development of Techniques

for Determining Distances

Classical Astronomical Techniques

Early astronomers estimated distances by comparing stellar brightnesses. They assumed that all stars are intrinsically equally luminous; hence, the brightest star would be the nearest. The study of double stars by William Herschel and others provided evidence that that simple assumption was not valid—intrinsic magnitudes covered a wide range. Consequently, new methods of estimating distances were devised.

One straightforward technique, known as trigonometric parallax, can be used to determine distances directly. This method was actually employed in 1838 by Friedrich Bessel to demonstrate that the Earth revolved around the Sun. Since the Earth does move around the Sun in its orbit, an observer on the Earth has a continually (but periodically) changing view of the heavens. Thus, the apparent positions of nearby stars will change with respect to more distant stars.

15

The amount of this shift in apparent position is inversely proportional to the distance to the star: nearby stars show a large shift, more distant stars show little or no shift. Unfortunately, this technique is applicable to only the nearest stars (usually less than 100 pcs) because the amount of the shift becomes vanishingly small for stars at even moderately large distances.

When astrometry—the measurement of stellar positions and motions—was refined in the nineteenth century, a technique based on proper motions, i.e., non-periodic motions of stars across the sky, became feasible. Although in practice, only small distances could be measured in this manner because proper motions for extremely distant stars are too small to be detected, the technique could extend the region of measurable distances beyond that available by the parallax method.

Around the Sun, stars move randomly in all directions with maximum velocities of only a few tens of kilometers per second. The local standard of rest is, by definition, the point at which the average velocity for all stars in the local vicinity is zero. Relative to the local standard of rest, the Sun is moving at about 20 km/sec towards the constellation Hercules. On the average, stars will reflect the Sun's motion by appearing to move 20 km/sec in a direction opposite to that of Hercules. Their apparent motions can be used to determine distances.

Imagine that we were in a car traveling at 60 mph (less than 0.1 percent the velocity of the Sun relative to the local standard of rest). Fence posts near the road would zip by, trees a few hundred yards distant would seem to pass more slowly, and mountains miles away would show almost imperceptible motion. The fence posts would have the greatest proper motion; i.e., the greatest apparent angular velocity. The greater the distance, the less the proper motion.

In a similar manner, we can determine the distance of groups of stars as long as we can measure their proper motion. Actually, since we are dealing with groups of moving stars, a better analogy would be a flock of birds circling near the road, a second flock across the field, and a third flock, which we would need binoculars to see, on the hilltops. Since in space we have no posts or hilltops, we must take group averages of proper motions.

The proper-motion technique was used to great advantage by several nineteenth-century astronomers. In particular, Jacobus C. Kapteyn used the method as a stepping stone to a more powerful technique. From the

16

proper motions available to him around 1900, Kapteyn[1] statistically derived the relative frequency of stellar magnitudes in the vicinity of the Sun. Then, with the assumption that the distribution of magnitudes was representative of distant regions of space, he calculated which fraction of stars at any given magnitude was

Jacobus C. Kapteyn: 1851–1922

J. C. Kapteyn was born in a small village of the Netherlands, and was one of fifteen children. His father, who ran a boarding school, first stimulated his interest in physics, which continued until he had obtained his doctorate. Although his formal training was not in astronomy, Kapteyn applied for and received the vacant position of observer at the Leiden Observatory. His work there won him a professorship, at the age of 27, at the University of Groningen. When he arrived at Groningen in 1887, he found no observatory. Although he was disappointed at first in his attempts to find funds for observational instruments, Kapteyn did not remain idle. He started work in conjunction with other observers, notably Sir David Gill at the Cape of Good Hope. Kapteyn busied himself with the analysis of photographic plates procured at the southern observatory.

At some time in his early life, Kapteyn had devoted himself to the task of determining the structure of the universe. His work in this field, for which he is most famous, continued up until his retirement, just before his death. Essential to his attack on the problem were vast amounts of data from which he could glean information by statistical analysis. To accomplish his goal, Kapteyn necessarily became a champion of international cooperation in astronomy.

In the year 1906, Kapteyn outlined a plan by which observatories all over the world would photograph stellar spectra and positions in specified regions of the sky. In this way the vast quantities of data necessary could be gathered rapidly. Unfortunately, World War I interfered tremendously with the spirit of cooperation. During the war, communications were slowed or severed; after the war, astronomers from the victorious Allied nations, especially from France, wished to ostracize astronomers from Germany, and even those from neutral countries. The Netherlands had been friendly to Germany, so that Kapteyn's pleas for understanding were suspect, even though he himself was friendly to the United States and had spent summers at Mount Wilson Observatory. Kapteyn argued that political acts of states did not represent the opinions of individual scientists, and that international cooperation in science was more important than petty vengeance. Despite his eloquent appeals, so many foreign astronomers refused to listen that international cooperation was impeded for many years.

intrinsically bright but distant, and which was intrinsically faint but near. Although the method could not provide a distance for an individual star, it could facilitate the calculation of the average spatial density of stars at various points in space.

The classical techniques of astrometry suffered from two serious deficiencies when applied to the estimation of distance. First, the method is applicable only to groups of stars and not to individuals. Second, it is restricted to relatively nearby stars. Even for the star-count method, with the greater sophistication added by Kapteyn, the basic data (the relative frequency of magnitudes) are firmly related to the nearby stars.

Henry Draper. (Wellesley College Observatory photograph)

Spectroscopic Techniques

Spectroscopy, which was applied to astronomical studies in the nineteenth century, held the key to many procedures for estimating distances. Although intrinsic stellar brightnesses vary considerably, many individual species of stars present similar characteristics, among which are temperature and luminosity. The latter can be used to estimate distances once the stars are classified as a given type from their spectra.

A useful classification scheme and a practical method of producing a catalogue of spectra resulted from a project begun in 1886 at the Harvard College Observatory when the widow of a physician-astronomer —Henry Draper—established a fund to support spectral studies.

Edward C. Pickering—the director of the observatory at the time—devised an ingenious method of producing spectrograms of many stars at once. He placed a narrow prism of low dispersion in front of the objective lens of his telescope; the lens focused the spectra on a photographic plate. In such a manner spectra of all the stars in the telescope's field of view were recorded: hundreds of stars yielded their spectra at one time.

With objective prism data, the Harvard group in 1897 completed the Draper Cataloque[2] in which the stars were classified according to their spectra, i.e., the types and intensities of various spectral lines. As a first attempt, the stars were classified according to the intensities of their hydrogen absorption lines, but that system of ordering soon presented problems.

18

When the stars were arranged so that their hydrogen lines formed a continuous sequence of intensities, the lines of all their other elements were discontinuous. Therefore, it seemed possible that something fundamental, affecting the spectra of all elements, not just hydrogen, underlay a spectral classification scheme. But did such a basic physical phenomenon actually exist? Could stars, in fact, be classified by their spectra? It is not intuitively obvious that the answer is yes. Annie J. Cannon, one of the Harvard workers, together with Pickering noticed that the spectral classes could be rearranged so that they formed a continuous sequence of gradual changes. When hydrogen had been used as a determinant, an alphabetical sequence had been adopted, with "A" stars being those with the strongest hydrogen line. Cannon's revised scheme scrambled the alphabetical ordering. Moreover, since bad spectra had been included in the original survey, fictitious classes had been created. Later these classes, such as "C" and "D", were included in the other spectra groups. Cannon's new classification system is the well-known system of spectral types used today—O, B, A, F, G, K, M, R, N, and S. [Henry Norris Russell suggested the useful mnemonic for remembering the sequence: *Oh, be a fine girl; kiss me right now, Sweet.*] Sample spectra of various classifications are shown. The

An early Brashear spectroscope used at Lick Observatory, *c.* 1887. (H. E. Mathews, San Francisco)

Edward Charles Pickering, director of Harvard College Observatory, 1876–1919. (Yale University photograph)

19

Annie Jump Cannon. (Sky Publishing Co. photograph)

Harvard investigators soon realized that the basic factor underlying the system was the surface temperatures of stars; O stars are the hottest, and the classification series represents a sequential gradual decrease in temperature.

While working on the Draper Catalogue, Antonia C. Maury, another Harvard researcher, noticed that the spectral lines of a given spectral type varied in width. Some stars had very broad lines, which Maury denoted with the letter "a." Those with very narrow spectral lines were denoted with the letter "c." Those stars that fell between, with lines of intermediate width, were denoted with the letter "b."[3] Maury's distinction between stars of a given spectral class was not included in the Draper catalogue, but her refinement was later used as an important tool in defining the scale of our Galaxy.

In 1905, Ejnar Hertzsprung found that the width of the spectral lines could be correlated with intrinsic brightness, by applying statistical methods to proper motions.[4] For a given spectral class, Hertzsprung found that Maury's c stars were brighter than a or b stars. Since all stars of a given spectral class have approximately the same surface temperature, Hertzsprung concluded that, in order for them to be so much brighter, those of type c must be geometrically larger than those of types a and b.

Maury, in noting the differences in line widths, had actually discovered a difference between giant and dwarf stars. Unfortunately, spectra were not well

Harvard's stellar classification team— known as "Pickering's harem." (Sky Publishing Co. photograph)

Ejnar Hertzsprung. (Sky Publishing Co. photograph)

understood at the time; the cause of the line broadening was unknown. Thus, Hertzsprung became the discoverer of giant stars. As happens so often in science, one worker made a discovery whose significance remained hidden until someone else put forward his own interpretations of the phenomenon.

Hertzsprung published his results in an obscure photography journal; consequently American astronomers were unaware of his discovery. Independently of Hertzsprung, Henry Norris Russell came to the same conclusions about the sizes of stars.[5] From photographs, Russell was able to determine parallaxes of stars, i.e., how far away they were. Once the distance d was known, the absolute magnitude M was easily calculated from the observed magnitude m from the equation

$$m - M = 5 \log(d) - 5.$$

Henry Norris Russell. (Sky Publishing Co. photograph)

From his calculations Russell concluded, as had Hertzsprung, that some stars within a spectral class are intrinsically much brighter than others. The two men differed, though, in their interpretations of the data. Russell believed in a two-stage process of evolution: stars were initially red giant stars that heated up while they contracted until they glowed blue-hot. During the second stage, the stars gradually cooled without contracting much further. During the first stage, the stellar brightness remained unchanged; during the second, the brightness decreased. Hertzsprung believed that the two stages were independent evolutionary paths.[6]

21

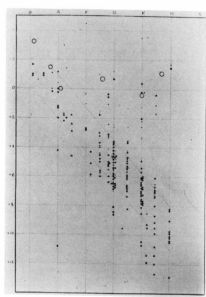

The Hertzsprung-Russell diagram as published by Russell in 1914. (*Popular Astronomy, 22,* 1914)

In 1911, Hertzsprung published a diagram of apparent magnitudes plotted against colors for stars in the Pleiades and Hyades clusters. In 1913, Russell published a similar diagram of absolute magnitude versus spectral class. Russell's plot is now known as the Hertzsprung-Russell, or H-R diagram. The plots drawn by both men showed clearly that the c stars are quite different from the rest. Because of their large size, indicated by high luminosities, the c stars are now called giant stars; the others are called dwarfs. As a result of Hertzsprung's and Russell's research, the distances of stars could be estimated if their spectra could be recorded. The spectra indicated the absolute magnitudes, which allowed simple calcuation of distances.

Cepheid Variable Stars

About the same time that Hertzsprung and Russell were discovering the differences between giant and dwarf stars, which was essential for distance estimates, Henrietta Swan Leavitt at the Harvard Observatory was laying the groundwork for another method of determining absolute magnitudes of a special class of stars.

In 1908, while studying variable stars (stars that periodically change in brightness) in the Magellanic Clouds—two large aggregates of stars seen in the southern sky—Leavitt[7] noticed that the longer the periods of the stars, the greater also were their luminosities. Four years later, she established that the relation between the logarithm of their periods and their apparent magnitudes was nearly linear.

The relation discovered by Leavitt had a potential for accurate determination of the distances of celestial objects, but she did not develop it. She recognized that her discovery could be used as an indicator of intrinsic brightness, but was prevented from pursuing the subject any further by Pickering, who believed their duty was to collect data, not interpret it.

The year following Leavitt's discovery, Hertzsprung[8] noted that the variables Leavitt had discovered in the Magellanic Clouds had light curves identical to those of Cepheid variables. Hertzsprung realized that if the Magellanic variables were indeed Cepheids, then by knowing the period of any Cepheid, its absolute luminosity could be determined, once Leavitt's period-luminosity function was calibrated. And, of course, from the absolute and apparent magnitudes the distance could be easily calculated. But

Henrietta Swan Leavitt, discoverer of the period-luminosity relation. (*Popular Astronomy, 30,* 1920)

calibration was no minor task. By chance, no Cepheid lies near enough to the Earth to allow its distance to be measured directly by the parallax technique; Hertzsprung therefore resorted to proper motion statistics. Leavitt's discovery, once calibrated, made possible distance determinations far beyond the limits of the parallactic method. Although the period-luminosity method was not a direct one, it was far more accurate and far more versatile than previously devised statistical methods, which depended on large numbers of stars for any accuracy; the Cepheid method of distance determination required only one Cepheid associated with a given object to be applicable.

The Large Magellanic Cloud photographed in 1903 from the Cape of Good Hope. (Royal Astronomical Society photograph)

Light curves of four Cepheid variable stars in Messier 31. (E.P. Hubble, *Realm of the Nebulae,* Yale University Press: New Haven, 1936)

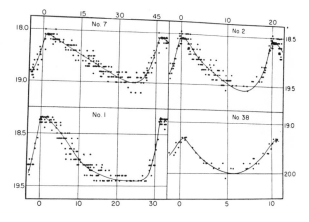

The Kapteyn Universe. Kapteyn's model of the Milky Way shown in cross section. The Sun's position is indicated above and to the right of the galactic center. (*Astrophysical Journal*, 55, 1922, University of Chicago Press)

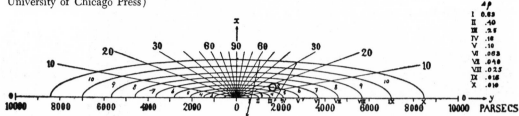

CHAPTER 3

Galactic Theories in the Period
1900-1920

The Kapteyn Universe

The techniques for determining the distances of stars and groups of stars allowed astronomers to begin to delve into the structure of our Milky Way Galaxy and to speculate on the relation of our Galaxy to the universe as a whole. How galactic research influenced cosmic theories will be discussed later. First, we shall examine the two major galactic models that were presented during the early twentieth century.

An attempt to determine the structure of the universe was made at the end of the nineteenth century. Just before 1900, there was a surge of new observational work that produced voluminous data; the Durchmusterung catalogue of stellar magnitudes and positions was completed. It included stars in the southern hemisphere, whose positions were determined by Jacobus C. Kapteyn from photographic plates.

Hugo von Seeliger soon determined the relative structure of the Galaxy by making counts of stars between successive magnitudes in various parts of the sky.[1] Von Seeliger's method, which gave the rates at which the Galaxy was thinning in various directions, eliminated the need for Herschel's assumption that we were able to see to the edge of the Galaxy. The flattened shape von Seeliger determined, however, was similar to Herschel's Galaxy, with the Sun located near the center.

24

In 1901, Kapteyn used statistics and proper motions of stars to determine the average distances to stars between successive magnitudes, and thus provided a scale for von Seeliger's work. The results of the investigations of these two men showed the Galaxy to be a flattened stellar system about 10 kpc in diameter and 2 kpc in thickness. Later, extensive studies by Kapteyn, using the method of star counts, confirmed his first general results for a model of the Galaxy.[2]

Kapteyn realized that his results were highly dependent upon an unverified assumption: that no absorption of light took place in space. If the amount of absorption were considerable, stars would appear dimmer, and therefore would seem to be farther than they actually are. Since the farthest stars would be affected the most by absorption, their apparent distances would suffer the greatest increase. The total effect would be to stretch the galactic model, the most distant parts being moved the most. Of course, the greater the apparent stretching, the lower the apparent density; hence interstellar absorption would make the Galaxy seem to thin out more rapidly than it actually does. Kapteyn realized that the existence of even moderate amounts of absorption would negate the conclusions of his analysis. He therefore initiated many research programs intended to measure the amount of absorption. Most of the results were inconclusive, and some indicated that if absorption did indeed exist, it certainly was not in sufficient amounts to seriously affect his galactic model. By 1918, Kapteyn was confident that the amount of interstellar absorption was negligible and that the "Kapteyn Universe"—as his model was often called—was at least a rough approximation to the structure of our Galaxy.

J.C. Kapteyn with his daughter. (Yale University photograph)

A class in practical astronomy at the University of California, c. 1895. (Yerkes Observatory photograph)

An 1887 photograph of the globular cluster Messier 13 taken with a 20-inch reflecting telescope. (I. Roberts, *Celestial Photographs*, Universal Press: London, 1893)

Shapley's Galactic Model

A young American astronomer—Harlow Shapley—arrived at the Mount Wilson Observatory in 1914 and began extensive studies of globular clusters. His research led him to a radically new galactic theory.

In 1915, Shapley[3] noted that globular clusters are widely distributed in galactic latitude symmetrically above and below the galactic equator, but they are largely concentrated into one hemisphere in galactic longitude. He also noted that this skewed distribution in longitude was unique to globular clusters; no other objects showed such a strange arrangement. In a footnote to a paper at that time, Shapley remarked that this asymmetry had "led Bohlin [1909] to suppose that they form a system at the center of the galactic system and the sun is eccentrically situated."[4] Since Bohlin's 1909 theory had been based upon very little sound evidence, it was not accepted. And in 1915, unable to reconcile the distances of globular clusters that he had calculated during his studies with the accepted values for galactic dimensions, Shapley also rejected Bohlin's idea.

The distances Shapley used for globular clusters were determined from the apparent magnitudes of certain types of stars. They included Cepheid variables, which Shapley could identify in the nearer globular clusters (typically 10 kpc away). Given the size of the Milky Way as determined by Kapteyn, the distances of globular clusters as derived with Cepheid variables indicated that they must lie outside the Galaxy. By 1916, Shap-

ley[5] had determined that Messier 13 (M13), a globular cluster in Hercules, was 30 kpc distant. That enormous distance placed M13 well beyond the boundaries of the galactic system as it was accepted by contemporary astronomers. On this point Shapley observed that "there seems to be reasonably clear evidence that M13 and similar globular clusters are very distant systems, distinct from our Galaxy and perhaps not greatly inferior to it in size."[6]

The nagging problem remained, however, of why globular clusters are distributed asymmetrically. If they are comparable in size to our Galaxy, yet physically removed from it, one would expect them to be distributed randomly, not asymmetrically. If they are outside our Galaxy but associated with it, their distribution might suggest that we—not they are in a skewed position!

A 1918 photograph of the globular cluster Messier 22 taken with the 60-inch telescope at Mount Wilson. (Hale Observatories photograph)

27

George Ellery Hale. (Hale
Observatories photograph)

Perhaps the Earth is located toward the edge of an enormous system, much larger than had been thought, which is defined by the globular clusters. The alternatives, whether the globular clusters are or are not associated with our Galaxy, presented a dilemma. If they are associated, why did Kapteyn find the Galaxy to be so much smaller? If they are not associated, how do we explain their distribution? These questions must have been considered by Shapley.

When more evidence had accumulated on the distribution of globular clusters, Shapley in 1917 reversed his position on Bohlin's interpretation, and boldly asserted that globular clusters are physically associated with our Galaxy. Moreover, he claimed that the Galaxy is actually ten times larger than astronomers had believed!

The reaction to these revolutionary ideas was mixed. George Ellery Hale, director of Mount Wilson Observatory gave Shapley cautious encouragement:

> . . . the outline given in [*Publ. Astron. Soc. Pacific*, 30, 1918] certainly seems to show that you have struck a trail of great promise. The distribution of the globular clusters with reference to the galactic plane certainly seems to indicate their organic connection with our own system, though the hypothesis regarding their dispersion in the "region of avoidance"* will need a lot of proof. However, I think you are right in making daring hypotheses . . . so long as you . . . are prepared to substitute new hypotheses for old ones as rapidly as the evidence may demand.[7]
> *[*This term denotes an area around the galactic plane devoid of globular clusters. Since loose clusters are found there, however, Shapley proposed that perhaps they were the remnants of globulars that had been tidally disrupted by the intense gravitational field of the Galaxy.*]

Notice that Hale only said Shapley's work "seems" valid. He clearly was not yet convinced. Many astronomers were far more doubtful than Hale.

Opposing Views and Island Universes

Shapley's theory met with strong opposition, for understandable reasons, most of which were related to the belief that spiral nebulae are galaxies similar to the Milky Way.

In 1914, even with the improved techniques for distance determination, the nature of spiral nebulae was unknown. In his annual report to the Lick Observatory, 15 May 1914, Heber D. Curtis[8] said that he planned to continue work on the exciting field of spirals because they were so perplexing. Recent studies had indicated they had unusually high radial velocities (see Section Three), and, as Curtis noted, they may be "inconceivably distant galaxies of stars or separate stellar universes so remote that an entire galaxy becomes but an unresolved haze of light." If the spiral nebulae were galaxies of the size proposed by Shapley, they would necessarily have been inconceivably distant. Since there was no evidence that they were so distant, many astronomers concluded that Shapley was wrong.

In addition, to the satisfaction of many astronomers, von Seeliger and Kapteyn had solved the related problem of the size and shape of our Galaxy.[9] It was generally believed that further refinements to the model would cause no major revisions.

One of the best distance indicators for spiral nebulae was discovered accidentally in 1917 by an astronomer at Mount Wilson, George W. Ritchey, who was making long-exposure photographs of nebulae in an attempt to measure their rotations and proper motions. In the course of his work, Ritchey discovered a nova in the spiral nebula N.G.C. 6946.[10] The object, discovered on a plate taken on 19 July 1917, was of apparent magnitude 15. Realizing that it was a nova, Ritchey began a search of Mount Wilson plates, and discovered several more novae.[11] Soon astronomers at other observatories were searching their plate collections, and within two months the number of known novae in spiral nebulae had increased to eleven.[12]

Curtis at Lick was one of the leaders in these discoveries. [Actually, Curtis[13] seems to have found a nova in a spiral three months before Ritchey discovered

Thirty-one globular clusters in the southern Milky Way. (H. Shapley, *Star Clusters*, McGraw-Hill: New York, 1930)

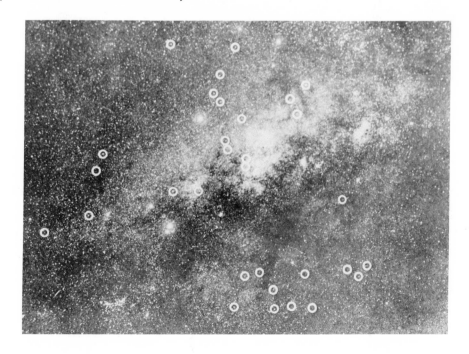

his first one. But since Ritchey brought his discovery to public attention first, in the literature he is usually credited with the finding.] For several years he had been photographing spirals with the 36-inch reflector, an undertaking similar to that started by James E. Keeler shortly before his death in 1900. Curtis hoped to compare the later photographs with the earlier ones to find evidence of motions of the spiral nebulae. Thus, when Ritchey at Mount Wilson telegraphed to Lick the news of his discovery, Curtis already had an excellent set of plates in which he could search for other novae. The nova on Ritchey's single plate might have been a coincidence, caused by the chance occurrence of a galactic nova in the line of sight to the spiral. As more novae were discovered, however, Curtis dispelled that possibility:

> It is possible that a single nova might appear, so placed in the sky as to be directly in line with a spiral nebula, tho [sic] the chances for such an occurrence would be very small. But that six new stars should happen to be situated in line with a nebula is manifestly beyond the bounds of probability; there can be no doubt that these novae were

George Ellery Hale: 1868–1938

There are few men who have had more influence on science than George Ellery Hale; his list of achievements is too long to include here. He established several observatories, among which are Yerkes, Mount Wilson, and Mount Palomar, as well as organizations devoted to science such as the National Research Council and the International Union for Cooperation in Solar Research. Furthermore, he carried out basic solar research that greatly advanced that branch of astronomy.

As a youngster, Hale was interested in science; his father encouraged him by buying microscopes and later a small telescope. A neighboring astronomer also encouraged Hale toward a career in astronomy. His father's real wish, however, was for Hale to become an engineer.

Four years of narrow technical training at the Massachusetts Institute of Technology confirmed Hale's decision that scientific research was of more interest to him than engineering, and he longed to return to the independent studies he had started years earlier. Fortunately, Hale was able to find part-time work at the Harvard Observatory, which cemented his interest in astronomy.

Hale was a man of considerable vision; he realized that the future of astronomy lay not in the measurement of stellar positions and magnitudes, but rather in applying the advances of physics and chemistry to astronomical

Andrew Carnegie (left) and George Ellery Hale at Mount Wilson. (Hale Observatories photograph)

actually in the spiral nebulae. The occurrence of these new stars in spirals must be regarded as having a very definite bearing on the "island universe" theory of the constitution of spiral nebulae.[14]

As this statement shows, Curtis regarded novae as the long-sought evidence that would finally settle whether or not spirals are complete stellar systems external to our own. By comparing the brightnesses of novae in spirals with those in the Milky Way, he obtained rough values for the distances of the spirals:

> There is . . . an average difference of 10 magnitudes between galactic novae and spiral novae. Now all evidence available assigns a great distance to the galactic novae. If we assume equality of absolute magnitudes for galactic and spiral novae, then the latter being apparently 10 magnitudes the fainter, are of the order of 100 times as far away as the former. That is, spirals containing the novae are far outside our stellar system; and these particular spirals are undoubtedly, judging from their comparatively great angular diameters, the nearer spirals.[15]

Curtis qualified his distance estimates by noting that if there exists absorbing material in the spirals, the novae

George Willis Ritchey, 1929. (Sky Publishing Co. photograph)

George Ellery Hale observing a solar image in the Great Hall of the National Academy of Sciences in Washington, D.C. (Hale Observatories photograph)

problems. He correctly prophesied that astrophysics would enable astronomers to reach out into space farther than ever before.

One of Hale's great contributions to science was his ability to organize and to raise funds. By appealing to the vanity of the country's millionaire tycoons. Hale was able to extract sufficient funds to finance the building of Yerkes Observatory, complete with a 40-inch refracting telescope —the world's largest at the time. Similarly, Hale was able to persuade other rich business men to finance larger telescopes.

A fascinating period of Hale's career started around 1904. Undertaking a tremendous gamble with his own money as well as with borrowed capital, he started work on an observatory on the top of Mount Wilson. In the beginning, conditions were primitive, but through hard work and considerable optimism, the project continued.

Eventually, in 1908, Hale had sufficient financial support to mount a 60-inch mirror. This marked the beginning of intense research into stellar and galactic structure. It was the 60-inch telescope at Mount Wilson that enabled Harlow Shapley to devise a new scheme for our Galaxy; with a later, even larger, telescope. Edwin Hubble was able to establish that spiral nebulae were galaxies like our own Milky Way. Although Hale's research was restricted mainly to the Sun, his organizational efforts bore fruit in many areas of astronomy.

The Mount Wilson Observatory

The idea of situating a telescope on a mountain peak was relatively new in 1900; most observatories were located at lower levels, near universities. George Ellery Hale was one of several who recognized the advantages of a mountain site, however, and in the year 1903 he decided to build a new research observatory on the top of Mount Wilson.

The only way of transporting equipment and materials was to use the backs of burros or mules. But through Hale's organization and determination, and with funds from the Carnegie Institute of Washington, telescopes, laboratories, and housing were erected. The conditions were nevertheless primitive, and water had to be hauled by hand from a nearby camp.

By 1909, a 60-inch reflecting teelscope had been mounted for use on the peak. Although funds had been provided for the mounting in 1904, many obstacles blocked completion of the project. The trail up the mountain had to be widened. A fire at the Union Iron Works in San Francisco during the 1906 earthquake nearly destroyed the mounting. Strikes delayed work on the dome for the telescope. Also, the mirror and the heavy steel mount had to be hauled behind mules up the narrow, tortuous trail. But by 7 December 1908, the 60-inch mirror was safely in its mounting.

The 60-inch telescope provided the instrumentation for many important discoveries. Hale's mind, however, envisioned an even larger instrument—a 100-inch reflecting telescope. With the promise of funds from John D. Hooker, a hardware magnate, Hale ordered a glass blank to be cast by the Saint Gobain glassworks in France. The blank arrived in Pasadena on 7 December 1908—the day the 60-inch mirror was secured in place on Mount Wilson—but the work of grinding the glass into the correct curvature did not begin for two years.

The Crossley telescope of Lick Observatory in 1898. (Lick Observatory photograph)

would appear dimmer; hence, farther away than they really are.

It should also be noted that Curtis, in his determination of the average magnitudes of novae in spirals, excluded two stars from his data; at their maxima, S Andromedae and Z Centauri both reached apparent magnitudes of about seven. Since these two novae were so much brighter than the rest, Curtis considered them anomalies.

At the same time that Curtis was using novae in spiral nebulae to confirm the island universe theory and to support Kapteyn's dimensions for the Milky Way, another Dutch astronomer and a student of Kapteyn—Adriaan van Maanen—was making measurements that were to be used as strong evidence against the galactic view for spirals.

The task of transporting the necessary materials and equipment for the 100-inch telescope was considerably more difficult than for the 60-inch; several near tragedies occurred. One time, a truck loaded with cement toppled over the edge of a 300-foot canyon. Two passengers jumped free; the driver did not, but luckily survived. A truck hauling part of the telescope tube also nearly slid over the edge of a precipice. Also, World War I interrupted the progress of the instrument. But finally, in November 1917, the telescope was tested and proved to be excellent.

The 100-inch could gather more than 2½ times the light of the 60-inch, and could see four times the volume of space. With the 100-inch, astronomers were able to establish that spiral nebulae were galaxies like our own, and that our universe seemed to be expanding. Although it has now been exceeded in size by several telescopes, in 1917 the 100-inch was a remarkable giant of great promise.

The ruins of the St. Gobain glass works in France resulting from the First World War. The 100-inch mirror was made here. (*Monthly Evening Sky Map*, June 1919)

In 1916, van Maanen published the results of extremely difficult measurements of proper motions of points in a spiral nebula.[16] His results (see Section Three) clearly indicated that the spiral was rotating. But if the angular rotation rate of the spiral were as great as van Maanen's study showed, and if it were as distant as Curtis' work suggested, then the physical velocity of its edges would have been prohibitively large. This conclusion follows from the fact that an object's proper motion μ is directly proportional to its tangential velocity v_T and inversely proportional to its distance d; i.e.,

$$\mu = \frac{v_T}{d}$$

Therefore, if both μ and d are large, as van Maanen and Curtis respectively claimed for spirals, then v_T must be extremely great.

The Crossley in its "new" mounting, c. 1910. (Lick Observatory photograph)

If van Maanen had measured only one spiral, his results might have been considered spurious by advocates of the island universe theory. But measurements made by van Maanen in later years on several spiral nebulae confirmed his initial finding. Moreover, even though his results were controversial, the evident care and precision of his work enhanced its credibility.

Shapley, who by 1917 had spent several years at Mount Wilson and had become a close friend of van Maanen, was more convinced than anyone else by van Maanen's results; he was certain that they discredited the island universe theory. In 1920, when Shaplely was forced to defend his theories vigorously, he was still strongly influenced by van Maanen's results even though, by then, other findings contradicted them. In 1967, while reminiscing about those days, Shapley remarked, "If you have large proper motion you are dealing with things near at hand . . . I went faithfully along with my friend van Maanen . . . although Curtis and Hubble and some others discredited van Maanen's measures and questioned his conclusions, I stood by van Maanen."[17]

As is often the case before full understanding is achieved, the observations at the time presented clear dilemmas: if van Maanen's measurements were valid, spirals must be close; if the results of novae studies were valid, spirals must be distant, even outside our Galaxy. If spirals were the size of the Milky Way as determined by Shapley, they would be farther than the distances determined from novae; if spirals were the size of the Milky Way as determined by Kapteyn, their distances would agree with the nova measurements.

Let us consider typical numbers. The spirals photographed in the first decades of this century typically have angular diameters of about 10'. If such a spiral had a linear diameter of 100 kpc, as Shapley later advocated for our Galaxy, its distance would be of the order of 30 Mpc. But if its linear diameter were 1/10th Shapley's value, its distance would be only about 3 Mpc. Curtis believed the novae in spirals to be roughly 100 times as distant as galactic novae, which occur at about 10 kpc; hence, to him; spirals would be at roughly 1 Mpc.

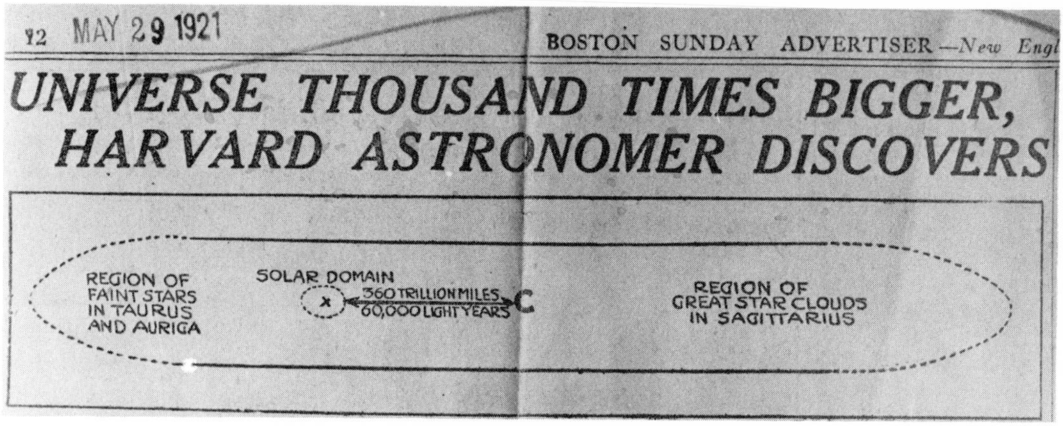

12 MAY 29 1921
BOSTON SUNDAY ADVERTISER—New Engl
UNIVERSE THOUSAND TIMES BIGGER, HARVARD ASTRONOMER DISCOVERS
REGION OF FAINT STARS IN TAURUS AND AURIGA
SOLAR DOMAIN
360 TRILLION MILES
60,000 LIGHT YEARS
C
REGION OF GREAT STAR CLOUDS IN SAGITTARIUS

A newspaper headline announcing Shapley's discovery of the large size of the Milky Way. (*Boston Sunday Advertiser*, 29 May 1921)

CHAPTER 4

Astronomy's Great Debate

Events Leading to the Great Debate

In Chapter 3 we saw that by 1920 several disagreements had arisen among scientists over almost all the basic questions of galactic astronomy. The fundamental source of much of that controversy centered on one of the most persistent and fundamental problems that man has faced—the accurate determination of distances. From that issue stemmed disagreements over numerous subjects, including the structure of the Milky Way and the nature of spiral nebulae.

With regard to spirals, Curtis had become the leading proponent of the old notion that they were island universes. But his arguments were different in kind from those of Kant and other philosophers; Curtis had solid evidence to support his claims. Because of Shapley's findings about the Milky Way and his belief in his friend van Maanen, he unwittingly became Curtis' chief critic.

With regard to the Milky Way, Curtis held to the view that the Galaxy's size had been correctly determined by Kapteyn and others from star counts. The new view, almost entirely due to Shapley, stated that the Milky Way is outlined in extent by globular clusters. Shapley's determination of distances for those objects revealed a galactic diameter of 100 kpc. This large diameter also ruled out the possibility that spirals were of a size comparable with our Galaxy. As discussed in the previous chapter, such a diameter, according to the island universe theory, would necessitate immense distances to the spirals, distances that were not compatible with observations. Consequently, Shapley's work on distances brought him into conflict with Curtis on two separate issues.

Several events occurred that led to a confrontation between Curtis and Shapley in 1920. The National Academy of Sciences was to hold its annual meeting at the Smithsonian Institution in Washington, D.C., on 26 April. In early 1920, Charles G. Abbot, Secretary of the Academy, began organizing a program of speakers. Hale proposed that a Hale Lecture, sponsored by a memorial fund in the name of his father, might occupy one evening.[1] Hale originally suggested that the lecture might take the form of a debate on the subject of relativity or perhaps island universes. Abbot believed that the new topic of relativity would be incomprehensible to the majority of the members of the Academy, who came from all fields of science.[2] He also feared that the audience might not be interested in a discussion of island universes; but, if that were to be the topic, he suggested W. W. Campbell, director of the Lick Observatory, and Shapley as opponents for such a debate.[3] Since the Academy was meeting shortly after the 1914–1918 war, the topic of medical progress in treating wounded soldiers was proposed as a reply to outcries from anti-vivisectionists. Eventually, as time was growing short, Abbot settled on the topic of the scale of the universe and chose Shapley and Curtis as the speakers.[4]

Shapley immediately accepted the invitation[5] but Curtis was reluctant to participate in a debate on the scale of the universe.[6] Since he was aware of Curtis' reluctance, and he himself was eager to present his views publicly, Shapley suggested that someone else be asked to take the opposing side.[7] Eventually, however, Curtis agreed to participate, as long as certain conditions were met about the procedure to be followed.[8]

Curtis agreed to the topic as first stated in Abbot's telegram[9]—the scale of the universe. Curtis interpreted "universe" to mean everything observable, including spiral nebulae. He objected to any title that would exclude spirals, and preferred the less precise title of "Scale of the Universe." In an outline of the proposed debate,[10] Curtis specifically included the spirals as part of the topic of discussion for both Shapley and himself.

36

Harlow Shapley and two of his children at the beach. (H. Shapley, *Through Rugged Ways to the Stars,* Scribners: New York 1969)

Harlow Shapley: 1885–1972

Harlow Shapley began his career as a crime reporter for a small-town newspaper in Kansas, covering the fights of drunken oil men. Later, he continued his rough occupation in Missouri.

Despite a sketchy education, Shapley was admitted to the University of Missouri when he decided to try higher education. He wished to study journalism, but when he arrived at the University that school had not yet been opened, and Shapley, by chance, began a career in astronomy. As Shapley describes what led to his decision, in his book *Through Rugged Ways to the Stars*: "I opened the catalogue of courses. . . . The very first course offered was a-r-c-h-a-e-o-l-o-g-y, and I couldn't pronounce it! . . . I turned over a page and saw a-s-t-r-o-n-o-m-y: I could pronounce that—and here I am."

After graduation, Shapley was encouraged by one of his professors to apply for a fellowship in astronomy at Princeton; he was chosen. At Princeton, he came under the influence of Henry Norris Russell, and gained professional maturity.

Shapley was offered a job at Mount Wilson in the year 1914. There, with the advantages of large telescopes, he was able to pursue a line of study suggested when he was still a graduate student—the study of globular clusters. For years he collected data on the globular clusters, working during the cold nights, until finally he evolved a radically new model of our Galaxy.

Fortunately for the progress of astronomy, Shapley was selfconfident, enjoyed a good argument, and was not at all conservative. His model differed so drastically from previously accepted views of our Galaxy that only a person with a strong personality would have been able to defend his opinions in the face of continued criticism. But in the end, Shapley prevailed: his galactic model was accepted by the astronomical community.

Shapley, however, wanted to discuss only the size of our own Galaxy; as it turned out, his treatment of spirals in the debate consisted of only three paragraphs.[11] Clearly, from the outset, the two men had divergent views not only on the scientific issues but also on what the subject of the debate was supposed to be. In recent years Shapley has argued that Curtis did not stick to the subject.[12] In fact, Curtis did stay on the subject, covering all aspects of it from our own Galaxy to distant spiral nebulae; but in doing so he emphasized spirals, his specialty, more than had been planned originally.

The difference in emphasis and direction of discussion of the two debators was to be expected. Shapley's interest for years had been globular clusters; consequently his attention was focused primarily on issues relating to them, such as the size and structure of our Galaxy. Curtis, on the other hand, had been photographing and studying spirals for years; hence his attention was focused on them. What research interest he felt for our Galaxy stemmed chiefly from his belief that it was a spiral itself. As might be expected, each man emphasized the areas in which he was most interested, and knew most about.

The meeting was not a debate in the usual sense; both Shapley and Curtis preferred the term "discussion." As originally suggested, the discussion was to consist of two talks,[13] first Shapley, then Curtis. But Curtis argued that such a program would be dull and confusing to the audience. He suggested that they chal-

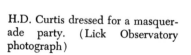

H.D. Curtis dressed for a masquerade party. (Lick Observatory photograph)

lenge each other's views, within the allotted time, so that the audience could judge the flaws as well as the merits of each theory. This plan was adopted.

The Debate's Importance

The confrontation between Shapley and Curtis in April 1920 was of great importance philosophically as well as scientifically. The issues at stake were those that the greatest intellects have pondered throughout history. What is the nature of the universe? The vastness of the space under discussion was (and still is) staggering.

The significance of the debate can be evaluated by comparing it with an hypothetical encounter between Copernicus and Ptolemy over their models for the solar system. Even though this analogy is instructive, for several reasons it is not precise. Copernicus and Ptolemy basically differed on only one topic: the arrangement of the solar system. By philosophy, religion, motivation, and approach, Copernicus was a conservative. His scientific terminology and constructions were almost identical to Ptolemy's. Even though his model carried profound implications, it was more a revision of Ptolemy's scheme than a fundamental shift in thinking.

Curtis and Shapley were opposed on two separate but related issues: the size of the Galaxy and the nature of spirals. Furthermore, to be more comparable to a debate between Copernicus and Ptolemy, the scale of the universe would have had to be argued between Shapley and Kapteyn, not Curtis. Kapteyn was responsible for the smaller galactic scale, as opposed to Shapley's diameter of 100 kpc. On this topic, Curtis was primarily the spokesman for and defender of the Kapteynian scheme.

Often in history the importance of an event is not recognized until time has elapsed and created a perspective. Such was the case in the Great Debate. Considering the impact the resolution of the debate was to have on philosophical as well as on scientific thinking, it is notable how little public interest was aroused. Indeed, at the time the debate was not even mentioned in any popular magazine in the United States or abroad. Brief articles about it did appear, however, in some newspapers. That the news media did not give the event wide coverage is perhaps understandable, but that the scientific journals should so ignore it is surprising. Apparently most of them failed to report it because of two factors. First, few scientists realized the event's significance; and second, the journals were devoted primarily, but certainly not exclusively, to research papers rather than to scientific reporting.

The reaction at the time of almost everyone, including the participants, was similar to that of the audiences at the maiden performances of Beethoven's *Eroica*, and Wagner's *Tristan und Isolde*—no one fully appreciated what he had heard.

The Debate

The contributions of Shapley and Curtis were somewhat out of character. Previously, Shapley had described his research in detailed, technical articles for scientific journals. But at the Academy meeting his presentation was given at a relatively elementary level. His reason for that approach was probably Abbot's[14] request that the papers be intelligible to a general audience and that details be left for publication. The audience was composed of scientists from all fields, most of whom would have found astronomical jargon unintelligible.

Curtis, who was experienced at speaking to general audiences, presented a comparatively technical paper. As he later admitted to Shapley,[15] his part of the discussion was too specialized for the audience. The reason for such a presentation is hard to understand unless, as Curtis[16] suggested in a letter to Shapley, he had expected Shapley also to present his side of the debate in a technical manner. In any case, the two sides of the argument were given on different levels.

Consequently Curtis and Shapley, with the urging of Hale, afterward decided to publish papers on the subjects of the debate. Since the evidence had not been presented in full, they decided that revisions of the original discussions should be published, and eventually decided on the *Bulletin of the National Research Council* as the logical place, since the Council had helped sponsor the meeting. Before the meeting, there had been little communication between Curtis and Shapley in preparing for the debate; for their joint publication, however, they freely exchanged views and working drafts of their papers.[17] This exchange enabled them to make rebuttals in the publication and to agree which issues to include and which to omit.

Shapley, whose arguments at the meeting had been cursory, went into elaborate detail in the published version, presenting all the technical arguments that favored his large size for the Milky Way. On the other hand, he still limited his discussion of spiral nebulae, to only three paragraphs. In his published version, Curtis kept fairly close to his original discussion, except that he added more specific details, especially in his attack on Shapley's theory.

Many important ideas were presented in the publication.[18] A critical, general point made by Shapley, which was basic to any further discussion of topics dealing with astronomical distances, was that the terrestrial laws of physics are valid everywhere in the universe. Unless this basic assumption is made, it is impossible to dicuss any event except those within our immediate neighborhood. There is evidence to support the assumption that physical laws are valid at all points in space. Perhaps the strongest evidence is provided by spectroscopy. Despite distances, the spectral lines of stars have the same spacings as those of terrestrial spectra, indicating that the physical elements behave in space exactly as they do on earth.

Other arguments presented were more technical; they depended upon data collected during the previous decade. Since the observations were not complete and were subject to varying interpretation, the two debaters were able to select the evidence that supported their views. The details of their discussion are presented in the following section.

The Evidence

Since the status of spiral nebulae and the size of the Milky Way Galaxy were discussed as two different, although related, topics, we shall present the evidence separately for each. In what follows, the evidence cited by Curtis appears on the right hand side; Shapley's views are found on the left. For clarity, their arguments have been arranged in four groups: Premises, Distances within the Milky Way Galaxy, Distances of spiral nebulae, and Conclusions.

According to Curtis, the most important of the distance arguments centered on the period-luminosity law of the Cepheid variables. It is interesting that Shapley made two mistakes concerning the variables, but his distance determinations were still basically correct. First, Shapley was wrong in assuming that the Cepheids in globular clusters do not differ from those in our neighborhood of the Galaxy; the nearer Cepheids are about four times brighter than those in the globular clusters. Second, Shapley underestimated the absolute brightness of the nearer variable stars by a factor of four. Consequently, by pure luck, his errors cancelled, so that his distances were essentially correct. His neglect of the dimming of stars due to absorption of light did, however, cause him to overestimate the distances to the globular clusters.

Curtis did err in his estimate of the brightness of galactic novae and the size of the Milky Way Galaxy; both of these estimates were too low. Consequently, he underestimated the distances of the spiral nebulae, but not so drastically as had Shapley. Shapley had been led astray by the research results of his friend Adriaan van Maanen. Van Maanen had been trying to measure the rotation rates of spirals and had succeeded in producing convincing results. Shapley naturally believed his friend's data, which implied that spirals were much closer than Curtis claimed.

As further research has shown, each debater was right in some ways and wrong in others. Shapley's revolutionary theory of our Galaxy displaced the conservative model defended by Curtis, yet Curtis prevailed in the realm of the spiral nebulae.

Several topics relevant to a discussion of the size and structure of our Galaxy were mentioned only briefly by the participants in the debate. One of these—interstellar absorption—was such an important issue that it is surprising a competent scientist would overlook it or ignore it intentionally.

The effect of absorption on distance estimates is especially crucial; in order for Shapley's relative distances to be correct, space would have to be transparent. If absorption exists, the errors in magnitude will vary with distance, the greatest error arising for the farthest stars. The effect of these errors will be to overestimate the distances, but the farthest objects will be affected the most.

Shapley had searched for the effects of selective absorption. He assumed that if there were absorbing material in space, it would have different effects on different wavelengths of light. He looked for such evidence in globular cluster stars.

Measurements of the difference in magnitude of a star, recorded on a photographic plate sensitive to blue light and on one sensitive to mean visual light, gave Shapley an indication of the star's color. (The difference between the two magnitudes is the color-index.) Shapley found that the distribution of color-indices for stars in globular clusters did not vary appreciably from cluster to cluster, even though the clusters seemed to be at greatly varying distances. If there were selective absorption (more absorption at one wavelength than at another), the distribution of color-indices would depend on distance. From these observations Shapley concluded that absorption of light in space is negligible. By assuming that his distances for globular clusters were accurate, Shapley[19] estimated selective absorp-

THE SCALE OF THE UNIVERSE

Shapley

Curtis

BASIC PREMISES

1. Globular clusters outline the extent of our Galaxy.
2. Stars in globular clusters are not peculiar, i.e., are similar to stars in our neighborhood.
3. Absorption of light is not a serious concern.

2. There is no evidence that stars in globular clusters are not peculiar.

DISTANCES WITHIN THE MILKY WAY GALAXY

1. Distances of many clusters can be derived by using the period-luminosity law. Average absolute magnitude of a typical Cepheid is about -2^m5.

1. Shapley's data are insufficient to be statistically accurate. He used only 11 stars to determine the average brightness. Furthermore, a large dispersion in the absolute magnitudes reduces the usefulness of the method for distance determination.

2. Hot stars (spectral type B) seen in globular clusters have an average absolute brightness of 0^m.
3. The 25 brightest stars in globular clusters are giant stars, and have an average absolute brightness of -1^m5.
4. Giant stars are the most prominent members of the globular clusters.

2. Hot stars (spectral type B) near the Sun have an average absolute brightness of 1^m6.
3. Giants have an absolute brightness of 1^m5.

4. Average stars have an absolute magnitude of about 5^m. Since the giants do not predominate in globular clusters, the distances are about 1/10 those determined by Shapley.

5. The physical diameters of globular clusters are approximately equal. Consequently, the distances of the clusters are inversely proportional to their angular diameters.

DISTANCES OF SPIRAL NEBULAE

1. The average brightness of novae in special nebulae as compared with novae in the Galaxy indicates distances greater than 150 kpc.
2. At a distance of 150 kpc, the Andromeda Nebula would have a diameter comparable to Kapteyn's value for the Milky Way Galaxy.
3. Spectra of spirals are compatible with the idea that spirals are similar to our Galaxy.
4. Large radial velocities indicate that spirals are not associated with our Galaxy. Furthermore, since their angular motion has not been detected, they must be very distant.
5. Edge-on photographs of spiral nebulae show bands of absorbing material. That material corresponds to the obscuring matter that must exist in our own Galaxy in order to produce the region of avoidance.

6. Measurements of the rotation of spiral nebulae by A. van Maanen, indicate that they must be much closer than Curtis' theory demands.

6. Star counts by Kapteyn and others indicate a galactic diameter of about 10 kpc.

CONCLUSIONS

1. The diameter of our Galaxy is about 100 kpc.
2. Spirals are not comparable in size with the Milky Way Galaxy and are relatively near.

1. The diameter of our Galaxy is about 10 kpc.
2. Spiral nebulae are galaxies like our own and range in distance from 150 kpc for the Andromeda Nebula to more than 3000 kpc for the farther ones.

tion to be less than 0.01 mag/kpc. Therefore, he rejected the larger values that had been calculated by Kapteyn.

Absorption of light in space played an important role in defining the size of our Galaxy. Unfortunately, Shapley underestimated the amount of interstellar absorption. As it turned out, the globular clusters he used to form his estimate were in the direction of relatively low space absorption. Other clusters, however, were subject to considerable interstellar absorption.

The neglect of absorption in the galactic plane introduced some difficulties with Shapley's theory. For instance, an important feature of the distribution of globular clusters, which Shapley noticed, is that they are not found in the Milky Way. At the time, the cluster known to be nearest to the plane was at a perpendicular distance from it of 1400 kpc. Shapley was unable to explain this peculiarity.

It is interesting to note that Shapley had found evidence of absorption but had ignored it. In April 1917, Curtis discovered what he thought was a spiral nebula within 2° of the galactic equator (actually 4° from the presently accepted equator) and less than 9° from the galactic center. Shapley soon realized that the object was a distant globular cluster. Because of the apparent magnitudes, the cluster appeared to be at about 70 kpc, corresponding to a parallax between 10 and 20 millionths of a second of arc. Yet its measured diameter corresponds to a parallax of approximately 25 millionths of a second of arc. The differences in distance measured by the two methods could be explained if one assumed the existence of absorbing material in the plane of the Galaxy.

Shapley realized that this discrepancy existed, remarking that "when we consider the magnitudes of the stars the angular diameter is larger than usual."[20] But since there is a definite dispersion in the diameter of globular clusters, the evidence does not conclusively prove the existence of absorption. Therefore, Shapley chose to assume that the diameter happened to be unusually large.

To explain the region of avoidance without the help of absorption, several years before the debate Shapley had proposed an imaginative scheme, which he did not reproduce in the debate. He suggested that globular clusters could not form as compact systems in the intense gravitational fields of the equatorial regions. Also, he proposed that any globular clusters that entered the equatorial region would become unstable and thus would be dispersed, forming the loose clusters found in abundance at low galactic latitudes. This last mechanism for eliminating globular clusters was necessary because of their high speeds. Travelling at speeds of several hundred km/sec, some globular clusters eventually would enter the equatorial region. But since no such cases are found, they must somehow disappear or be destroyed.

Curtis, on the other hand, had a more reasonable explanation, based on photographs of spirals that showed dark lanes lying in the planes of greatest extension. An outer ring of occulting matter is particularly conspicuous in spirals photographed edge-on. Curtis reasoned that if our Galaxy is a spiral, it would have occulting matter in its plane. That obscuring matter would block faint objects, so that even though spirals are distributed randomly, those in the plane would not be visible. Thus, by assuming our Galaxy to be a spiral, Curtis was able to explain the region of avoidance. Indeed, evidence for clouds of obscuring matter had been cited from studies of regions in the Milky Way where stars appeared to be absent. The reasoning was slightly circular, however. Curtis was aware of this flaw in the logic, but he considered the visual evidence to be strongly in favor of the island universe theory.

Aftermath of the Debate

The debate did not end the controversy; each participant held firmly to his original view and the scientific community remained divided. Curtis was sure he had won the debate when he wrote that the "debate went off fine in Washington, and I have been assured that I came out considerably in front."[21]

Shapley disagreed and in his later years stated, "I think I won the debate from the standpoint of the assigned subject matter.[22] Although astronomers remained divided into two camps, the debate did clarify many of the arguments. Unfortunately, research in crucial fields had not progressed sufficiently for either Curtis or Shapley to reconcile their differences.

On April 21, 1921, a year after the debate, William H. Pickering (the brother of E. C. Pickering, who had been the director of the Harvard College Observatory) published a lengthy article entitled "The Dimension of our Stellar System."[23] In that remarkable paper, he reviewed the current observations and theories on galactic astronomy, but he did not mention the debate. He stated that "Shapley assumes that the center of the galaxy and the center of the globular clusters coincide. This seems natural and plausible, but he makes no attempt to prove it. Probably it can not be proved."[24] From an almost geometric calculation (assuming the Galaxy to be the shape of a double-wedge) he concluded that the diameter of the Milky Way is about 10 kpc and that the Sun is about 3800 kpc from the center. After accepting Shapley's basic structure as correct, with the Sun eccentrically located, Pickering calculated distances that are reminiscent of Curtis.' Pickering's review was, in essence, a compromise between two competing systems.

A contemporary review of the debate itself was presented by Peter Doig, a well-known historian of astronomy, at a meeting of the British Astronomical Association on 28 December 1921; later it was rearranged and published.[25] Apparently, over a year and a half elapsed after the debate before this single scientific review of it appeared.

Doig gave an itemized accounting of the arguments; then, as a critic, he stated that "The case for the larger scale of distances seems distinctly the stronger one." He went on to explain his reasons for that conclusion, but he failed to state his opinion regarding Shapley's eccentric position for the Sun, based on globular clusters.

Hector Macpherson, in an account of William Herschel's "World-views in the light of modern astronomy," claimed that

Shapley has undoubtedly confirmed the later Herschelian theories about the extent of the galactic system. A diameter of at least 300,000 light-years along the plane is obviously demanded by the facts which Dr. Shapley has brought together; and although his conclusions have been contested by several astronomers, notably by Dr. Curtis, the evidence in their favour appears to be overwhelming.[26]

He then described Shapley's views on the eccentric position of the Sun, but he did not state explicitly whether he agreed or not. He had just stated that Shapley's conclusions (on distances at least) were definitely correct, but in an article that was so uniform in praise of Herschel, it seems doubtful that Macpherson would have supported Shapley's galactocentric views, which contradicted the results of star counts. Perhaps Macpherson was torn between conflicting loyalties. He had corresponded extensively with Shapley about atronomical matters and a friendship had developed; nevertheless, Macpherson was obviously dedicated to the grand historical image of Herschel as England's greatest astonomer. In an apparent attempt to compromise, Macpherson construed Shapley's results as complementary to Herschel's, although in fact they

46

were contradictory.

However, the views of the three individuals just cited—all of whom at least supported Shapley's distance determinations—were far from universal at the time. Many, perhaps most, astronomers still held to the star count method. Kapteyn and van Rhijn in 1920 presented a provisional derivation of the star density in the Milky Way. These studies were expanded and resulted in 1922 in their famous "First Attempt at a Theory of the Arrangement and Motion of the Sidereal System."[27] Using star counts, Kapteyn deduced a model for our stellar system that was similar to his earlier model—the Galaxy was pictured as an enormous cluster of stars, but relatively small in extent. This flattened ellipsoid of revolution was centered approximately (but not exactly) at the Sun. For an inexplicable reason, neither Shapley nor Curtis were mentioned in that paper.

Kapteyn and van Rhijn[28] later criticized Shapley because they believed that the high proper motions of Cepheid stars showed them to be fairly close; hence, those stars should be dwarfs, not giants as Shapley had supposed. Today we know that those stars possess high space velocities and actually are distant. Thus, their high proper motions arise from their speeds rather than their distances.

Clearly, the Galaxy and spirals were almost as perplexing during the 1920's as they had been in the previous decade. Reviews of the evidence tended to reflect each author's opinion; in no way, except perhaps psychologically, did it support conclusively the Kapteynian system for the Galaxy.

H. Shapley's graph relating parallax and angular diameter for globular clusters. (*Astrophysical Journal, 48,* 1918, University of Chicago Press)

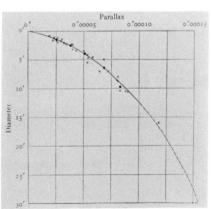

Epilogue

for Section One

In Section I, our investigation of the universe has evolved along two complementary tracks—study of the Galaxy and of spiral nebulae. At times, progress in understanding these two subjects seemed to proceed independently of one another; however, at other times, the two were intimately related. The two areas finally converged at the "Great Debate."

Soon after the confrontation between Shapley and Curtis, the two fields of investigation again diverged. Hubble's discovery of Cepheids in spirals definitely placed those nebulae outside the Milky Way, showing that they were worthy of study in their own right.

Sections II and III follow this historical and logical separation of the topics. Section II treats the problems and theories encountered in galactic astronomy, including the resolution of the conflict between Shapley's galactocentric model and Kapteyn's heliocentric model and, therefore, the development of a detailed scheme for the Galaxy. Section III continues the study of extragalactic nebulae. The conflict between the measurements of Hubble and of van Maanen is resolved and, from the study of extragalactic systems, fantastic cosmological implications begin to emerge. By continuing the study of the two subdivisions of the Great Debate, Sections II and III explore both the near and far realms of the universe.

References

Prologue

1 O. Struve and V. Zebergs, *Astronomy of the 20th Century* (New York: Macmillan, 1962), 416.
2 P. Van De Kamp, *Publ. Astron. Soc. Pac.* 77 (1965): 325.
3 G. Abell, *Exploration of the Universe*, 2nd ed. (New York: Holt Rinehart and Winston, 1969), 608.

Chapter 1

1 M. K. Munitz, ed. *Theories of the Universe* (Glencoe, Illinois: The Free Press, 1957), 230.
2 T. Wright, *Second or Singular Thoughts Upon the Theory of the Universe*, ed. M. A. Hoskin (London: Dawson's of Pall Mall, 1968).
3 *Ibid.*, 25.
4 *Ibid.*, 27.
5 *Ibid.*, 28-29.
6 Munitz, *Theories*, 231.
7 *Ibid.*, 234.
8 *Ibid.*, 237.
9 M. A. Hoskin, *William Herschel and the Construction of the Heavens* (New York: W. W. Horton and Co., Inc., 1964).
10 *Ibid.*, 99.
11 *Ibid.*, 38.
12 A. Berry, *A Short History of Astronomy* (reprinted by (New York: Dover Publications, Inc., 1066), 400. Original publication (London: J. Murray, 1898).

Chapter 2

1 O. Struve and V. Zebergs, *Astronomy of the 20th Century* (New York: Macmillan, 1962), 190 ff.
2 "The Draper Catalogue of Stellar Spectra," *Harvard College Observatory Annals* 27 (1890). The Catalogue was begun in 1886. The first part was published in 1890, and an additional volume including the stars of the Southern hemisphere appeared in 1897. The entire catalogue was revised in accord with the modern stellar classification system during 1918-1924: A. J. Cannon and E. C. Pickering, "The Henry Draper Catalogue," *Harvard College Observatory Annals 91–99* (1918–1924).
3 O. Struve and V. Zebergs, 195.
4 *Ibid.*, 195–196.
5 *Ibid.*, 197.
6 *Ibid.*, 200.
7 H. S. Leavitt, "1777 Variables in the Magellanic Clouds," *Harvard College Observatory Annals 60* (1908), 87–108.
8 H. Shapley, ed. *Source Book in Astronomy 1900–1950* (Cambridge: Harvard University Press, 1966), 253.

Chapter 3

1 R. L. Waterfield, *A Hundred Years of Astronomy* (New York: Macmillan, 1938), 127.
2 J. C. Kapteyn, "First Attempt at a Theory of the Arrangement and Motion of the Sidereal System," *Astrophys. J.* 55 (1922): 302–327.
3 H. Shapley, "The General Problem of Clusters," *Mt. Wilson Contr. 115* (1915).
4 H. Shapley, "Thirteen Hundred Stars in the Hercules Cluster (M13)," *Mt. Wilson Contr. 116* (1916): 86.
5 H. Shapley, "Outline and Summary of a Study of Magnitudes in the Globular Cluster Messier 13," *Publ. Astron. Soc. Pac.* 28 (1916): 174.
6 *Ibid.*, 176.
7 Private communication, G. E. Hale to H. Shapley, 14 March 1918 (Harvard University Archives).
8 H. D. Curtis, Progress Report for Mount Hamilton, 1 July 1913–15 May 1914.
9 O. Struve and V. Zebergs, *Astronomy of the 20th Century* (New York: Macmillan, 1962), 411.
10 G. W. Ritchey, "Novae in Spiral Nebulae," *Publ. Astron. Soc. Pac.* 29 (1917): 210.
11 G. W. Ritchey, "Another Faint Nova in the Andromeda Nebula," *Publ. Astron. Soc. Pac.* 29 (1917): 257.
12 H. D. Curtis, "New Stars in Spiral Nebulae," *Publ. Astron. Soc. Pac.* 29 (1917): 180.
13 Private notes, H. D. Curtis, 1917 (Mount Hamilton Archives).
14 Curtis, "New Stars," 181–182.
15 H. D. Curtis, "Novae in Spiral Nebulae and the Island Universe Theory," *Publ. Astron. Soc. Pac.* 29 (1917): 207.
16 A. van Maanen, "Internal Motion for Spiral Nebula Messier 101," *Astrophys. J.* 44 (1916): 210.
17 H. Shapley, *Through Rugged Ways to the Stars* (New York: Scribners, 1969), 80.

Chapter 4

1 Private communication, C. Abbot to G. E. Hale, 3 January 1920 (NAS-NRC Archives).
2 *Ibid.*
3 *Ibid.*
4 Private communication, C. Abbot to G. E. Hale, 18 February 1920 (Hale Collection).
5 Private communication, H. Shapley to G. E. Hale, 19 February 1920 (Hale Collection).
6 Private communication, H. D. Curtis to G. E. Hale, 20 February 1920 (Hale Collection).
7 Shapley to Hale, 19 February 1920.
8 Private communication, H. D. Curtis to G. E. Hale, 26 February 1920 (Hale Collection).
9 Abbot to Hale, 18 February 1920.
10 Curtis to Hale, 20 February 1920.

11 H. Shapley, Debate Manuscript (Harvard University Archives).

12 H. Shapley, *Through Rugged Ways to the Stars* (New York: Scribners, 1969) 79.

13 Private communication, H. D. Curtis to H. Shapley, 26 February 1920 (Hale Collection).

14 Private communication, C. Abbot to NAS members, 12 January 1920 (Hale Collection).

15 Private communication, H. D. Curtis to H. Shapley, 13 June 1920 (Harvard University Archives).

16 *Ibid.*

17 Private communication, H. D. Curtis to H. Shapley, 2 August 1920 (Harvard University Archives).

18 H. Shapley and H. D. Curtis, "The Scale of the Universe," *Bull. Nat. Res. Coun. 2*, pt 3, no. 11 (May 1921): 171.

19 H. Shapley, "Globular Clusters and the Structure of the Galactic System," *Publ. Astron. Soc. Pac. 30* (1918): 42–54.

20 H. Shapley, "Notes on the Distant Cluster NGC 6440," *Publ. Astron. Soc. Pac. 30* (1918): 253.

21 Private communication, H. D. Curtis to his family, 15 May 1920 (Allegheny Observatory Archives).

22 H. Shapley *Through Rugged Ways to the Stars*, 79.

23 W. H. Pickering, "The Dimensions of our Stellar System," *Publ. Astron. Soc. Pac. 32* (1921): 140.

24 *Ibid.*, 156.

25 P. Doig, "The Scale of the Universe," *J. Brit. Astron. Assoc. 32* (1922): 111.

26 H. Macpherson, "Herschel's World-View in Light of Modern Astronomy," *Observatory 45* (1922): 259.

27 J. C. Kapteyn, "First Attempt at a Theory of the Arrangement and Motion of the Sidereal System," *Astrophys. J. 55* (1922): 302–327.

28 J. C. Kapteyn and P. J. van Rhijn, "The Proper Motions of Cepheid Stars and the Distances of the Globular Clusters," *Bull Astron. Soc. Neth. 1* (1922): 37.

TWO

Galactic
Astronomy

Contents

of Section Two

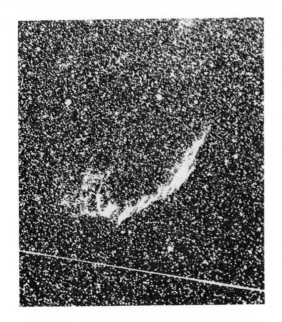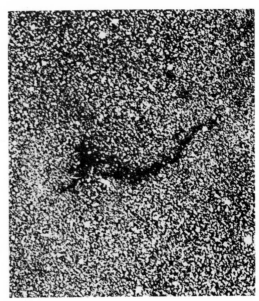

Two types of nebulae: (left) an emission nebula; (right) an absorption nebula. Note the similarity of shape. (Yerkes Observatory photograph)

Prologue

for Section Two

To some extent at least, people have always been concerned about their environment. For a sizeable fraction of them, this concern centered on their immediate surroundings. By necessity they had to know where to find water and game and fruit during the various seasons; the demands for existence itself did not allow time for exploration outside the immediate surroundings.

As the amount of leisure time for thought increased and civilization progressed, horizons expanded. But even at the peak of the Greek era the prevailing viewpoint was parochial. The Greeks were able to conceive of a three-dimensional space containing the planets on concentric spheres, but that space was bounded by the celestial sphere upon which all the stars were supposedly fastened. The vicinity of the earth—the solar system—was visualized in three dimensions, whereas the stars were pictured as being on the two-dimensional surface of a sphere.

Possibly the first description of the distribution of the stars that remotely resembled the modern view was expressed in the middle of the eighteenth century when Thomas Wright of Durham proposed a model to explain the Milky Way. He suggested that the stars were scattered randomly between two concentric planes, forming a thin shell. If we were located near the middle of the shell, Wright explained, we would see more stars in the direction of the plane tangent to the shell than in the direction perpendicular to it. The large number of stars we would see towards the tangent plane would be the Milky Way.

Five years later Immanuel Kant expounded the same general theory, reasoning by analogy from the solar system to the Galaxy. The planets all move about a central point in nearly circular orbits in nearly the same plane. Similarly, Kant believed, the stars must revolve about a common center in a common plane. This model would yield an apparent configuration similar to that which Wright had proposed, except that it would contain motion. And continuing his reasoning by analogy with planets and stars, Kant proposed that there must be a multiplicity of galaxies, each similar to our own.

The theories of Wright and of Kant were based on simple observation and much speculation. The first really scientific attempt to understand the structure of the Galaxy—one based on numerous, detailed observations—was undertaken by William Herschel.

Herschel immigrated to England in 1757 from Hanover, Germany, where he had been an oboist in the Hanover Guards. He was a musician by trade but spent his spare time studying the heavens. While doing so, on 13 March 1781, he accidentally discovered a faint, fuzzy-looking, greenish object in the sky. At first he thought it a comet or nebulous star, but soon realized that he had found a previously undiscovered planet. To add to the drama of the event, Frederick the Great is reputed to have stated in about 1780 that everything of importance in science had already been discovered.

In honor of the monarch of his new country, Herschel proposed naming the planet after King George III. Even though it was ultimately named Uranus, Herschel received a royal grant that enabled him to spend all of his time on astronomical pursuits.

The discovery of Uranus was a fortunate event for Herschel in several ways. First, through the patronage of George III, Herschel was able to continue his astronomical studies full time. Also, the prestige acquired by the discovery enabled Herschel to receive recognition by the scientific societies of the day.

One area of study that occupied Herschel most of his life was the determination of the structure of the Milky Way, our Galaxy. The method he used was rather simple. Using his large telescope, he counted the number of stars he could see in a given direction and related that number to the distance to the edge of the Galaxy. This method assumed that all stars were approximately of the same brightness and were uniformly spaced; that the telescope could see to the very edge of the Galaxy; and that absorption of starlight by an interstellar medium was negligible. Even though none of these assumptions was correct, Herschel at least took the first step toward determining galactic structure.

Later, in the twentieth century, more refined methods of taking star counts indicated a Galaxy roughly the same as Herschel's but much more symmetrical. Also, a more nearly spherical and much larger galactic structure was proposed, based on the study of certain clusters of stars. Even though these star clusters (called globular clusters) were separated from each other by vast regions of space, they outlined an ellipsoid centered on the Milky Way. The view that our Galaxy was outlined by the globular clusters opposed the view based on star counts. The story of the rise of the two views, the resolution of the conflict, and a detailed view of the structure of our Galaxy follows in this Section.

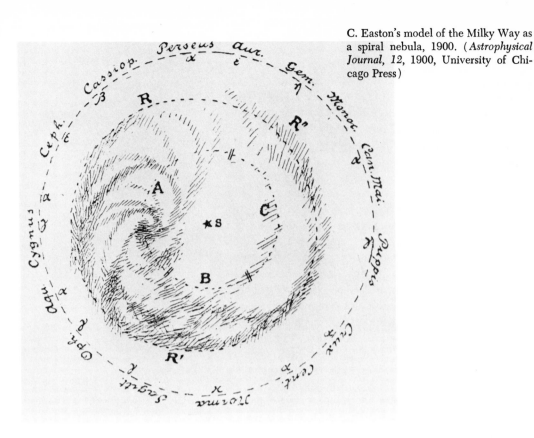

C. Easton's model of the Milky Way as a spiral nebula, 1900. (*Astrophysical Journal*, 12, 1900, University of Chicago Press)

CHAPTER 1

Galactic Models,

1900-1920

C. Easton's Spiral Model

Models for our Galaxy were devised as early as the mid-eighteenth century by philosophers, but William Herschel—the insightful English astronomer famous for discovering the planet Uranus—was the first to apply mathematical analysis to astronomical observations to deduce a model of our Milky Way. Herschel concluded that our Galaxy was elongated in the direction of the Milky Way and possessed an irregular outline. Although there were many faults in his analysis, Herschel's model provided the first approximation to the shape of our galactic system.

John Herschel—William's son—is credited[1] with being the first astronomer to suggest that the elongation of our Galaxy is due to rotation of the system (the flattening of a fluid or elastic body upon rapid rotation is a well-known phenomenon). His common-sense reasoning was correct, but many astronomers ignored the effects of rotation.

Cornelis Easton[2], around 1900, proposed that our Galaxy was a spiral nebula. He postulated that the appearance of the Milky Way could be accounted for by spiral structure. Recent photographs of spiral nebulae gave encouragement to his speculations and he produced a rough sketch showing the location of the Sun in a very distorted spiral. He qualified the sketch with the statement, "I wish to insist upon the fact that the sketch *does not pretend to give an even approximate representation of the Milky Way*"[3] (Easton's italics).

In 1913, he published an article[4] in which he made some astute conjectures. Without proof, he assumed that the Milky Way was a spiral nebula, like those photographed by George Ritchey and others, and asked if the appearance of the sky could be explained by the spiral structure. He found that by locating the spiral arms in different planes and placing the Sun in an arm about halfway from the center to the edge, he could make his model agree with the observations. In fact, encouraged by analogies to spiral nebulae, Easton felt confident he was on the right track. Compared with modern theories Easton's guess, even to the position of the Sun, was not far wrong.

An 1889 photograph of the spiral nebula Messier 51 taken with a 20-inch reflecting telescope. (I. Roberts, *Celestial Photographs*, Universal Press: London, 1893)

Hugo Seeliger.
(Yerkes Observatory photograph)

Easton made another good guess. In the sky of the southern hemisphere there are luminous patches known as the Magellanic Clouds. No one was quite certain what these star clouds were or what their relation to the Milky Way was. Easton reasoned that "it is natural to think that the Magellanic clouds should be analogous to the small nebula connected with Lord Rosse's spiral M51."[5] We now know that the Magellanic Clouds are gravitationally connected to the Milky Way Galaxy.

But not all of Easton's surmises were correct. He did not think that the smaller spiral nebulae were external galaxies, and refused to admit that the existence of the small spirals had any bearing on his theory: "I wish only to remark that nobody will deny the existence of a whirlpool because he sees a number of small eddies in the convolutions of the great one."[6]

Models derived from Star Counts

One of the first people after Herschel to use star counts to find the general shape of our Galaxy was Hugo von Seeliger of Munich. He noticed that the number of stars did not increase with increasing magnitude by the factor to be expected if the number of stars per unit volume, i.e., star-density, were constant. For a change of one magnitude, the limiting distance to which we can see stars increases by $\sqrt{2.51}$; i.e., the square root of the ratio of luminosities differing by one magnitude. The volume up to the limiting distance should increase by a factor of $(\sqrt{2.51})^3$, or approximately four. Hence, for a change of one magnitude in the limiting brightness, one would expect to see four times as many stars—if the star-density were constant. Von Seeliger found the factor to be about three.[7] He interpreted this discovery as a decrease in the galactic density with the distance from the Sun. The decrease was more rapid toward the galactic poles than toward the Milky Way, which indicated that our Galaxy is elongated in the direction of the Milky Way.

By 1900, and during subsequent years, the Dutch astronomer Jacobus Kapteyn was also attempting to determine the structure of our Galaxy from star counts. To extract a more realistic model than that of von Seeliger, Kapteyn had to assume that, although the number of stars per absolute brightness in space was not known, the relative frequency of absolute magnitudes was the same throughout the rest of space as among the local stars. As a result of his studies, Kapteyn found that our Galaxy was similar in shape to

58

that determined by von Seeliger. Kapteyn had been precise in his work and was confident of the results, but he realized the need for more and better data to increase the statistical accuracy of his methods of studying galactic structure. Hence, in 1904, he proposed his famous "Plan of Selected Areas."[8] His plan called for the cooperation of observatories throughout the world in photographing the stars in 206 specific areas distributed throughout the sky. Each cooperating observatory was assigned a section of sky in which to determine the magnitudes, colors, proper motions, radial velocities, and spectral types of all stars brighter than a certain limiting magnitude. This extensive international effort produced the data that Kapteyn used to construct a definitive model of our Galaxy.

Kapteyn needed international cooperation to reach his goal for two reasons. First, the inclement weather of the Netherlands forced the astronomers there either to travel, at great cost in money, time, and comfort, to other observatories, or to rely on other persons' data. Second, the amount of data required was too vast to be collected by one or even a few astronomers within a reasonable time span; many observers were needed.

J.C. Kapteyn.
(Yerkes Observatory photograph)

A detailed analysis[9] of the stellar distribution of our Galaxy did not appear in published form until 1922, the year of Kapteyn's death; but he privately communicated his preliminary results to other astronomers before then. His model, which became known as the "Kapteyn Universe," was ellipsoidal with an axial ratio of 5:1 and a major axis measuring about 16 kpc. The density gradually decreased with distance from the center, so that the density at about 8 kpc was one-hundredth the value in the solar neighborhood.

The most undesirable feature of Kapteyn's model was the location of the Sun. The calculations revealed a nearly central position, which led some astronomers to suspect the validity of the model—the Copernican anti-anthropocentric tradition argued against such a favored location. Unfortunately, Kapteyn derived his model on the simplifying assumption that the Sun was located at the center, which led some astronomers to criticize him on philosophical grounds. The truth is that Kapteyn emphasized the more detailed result of the Sun's non-central position: ". . . *the position of the center of the system* is not the sun, as assumed, but is probably located at a point some 650 parsecs away in the direction galactic long. 77°, lat. −3°. . . ."[10] (Kapteyn's italics). Nevertheless, the near central position assigned to the Sun raised doubts in many minds, including Kapteyn's.

Although Kapteyn was certain that his calculations indicated an apparent decrease in the density of the stellar distribution with distance from the Sun, he realized that an explanation other than the "Kapteyn Universe" was possible:

. . . we may conclude that if there is a thinning-out of the stars for an increased distance from the sun, it must be so in whatever direction from the sun we proceed. This would assign to our sun a very exceptional place in the stellar system, viz., the place of maximum density.

On the other hand, if we assume that the thinning-out of the stars is simply apparent and due to absorption of light, the apparent thinning-out in any arbitrary direction is perfectly natural.[11]

Since he was concerned with the structure of our galactic system, Kapteyn was anxious to know if the effect described above was real or merely apparent. If it were real, we were near the center of the system; unless absorption occurred, no other deduction was reasonable.

As we shall discuss in more detail later, Kapteyn searched for absorption and did not find any in sufficient quantity to suggest that the apparent "thinning-out" of stars was not real. Consequently, he was forced to the conclusion that our Galaxy was ellipsoidal and about 16 kpc in diameter.

Shapley's Model Based on the Distribution of Globular Clusters

As we discussed in Section One, Harlow Shapley,[12] during a study of the distribution of globular clusters, came to the conclusion that the clusters were physically associated with our Galaxy and that their distances indicated the extent of the Milky Way system. His distances, based primarily on the period-luminosity relation of Cepheid variable stars, were highly discordant with the galactic diameter deduced by Kapteyn. In fact, Shapley's estimate of the size of our Galaxy was about ten times that of Kapteyn. Naturally, the discrepancy between the two models caused great concern among astronomers.

There were several reasons for this concern. First, there was the desire to know the structure of our local system. Perhaps more important, there was the desire to know our place in the structure of the cosmos. Was our solar system located far from the center of a large stellar system, or near the center of a small one? Was our Galaxy unique in the universe, as Shapley's results indicated, or was it similar to the spiral nebulae seen scattered throughout space? These questions were pressing around 1920 when the "Great Debate" discussed in Section One took place.

The answers to the cosmological questions posed above lay in the accurate determination of distances. Kapteyn's model rested on solid astronomical foundations since it was based on the easily ascertained relative distances of stars. The scale of his system could then be derived from the distances of nearby stars. These distances were accurately established. Shapley's distances were vast and, hence, were not directly attained. Shapley had to proceed from one technique to another as the distances became greater. The distances of the farthest clusters were determined as relative to nearer clusters by comparison of apparent diameters. The distances of the nearer clusters were deduced from the apparent magnitudes of stars thought to be typical of certain kinds of stars. The uncertainty lay in two areas. First, did the stars chosen by Shapley belong to the class assumed? Second, were the assumed absolute magnitudes of the classes correct? Because of these uncertainties, some astronomers were reluctant to accept Shapley's model.

A 1909 photograph by G.W. Ritchey (Mount Wilson) of the supernova remnant known as the Crab Nebula. (Hale Observatories photograph)

CHAPTER 2

Galactic Rotation

Enigmatic Developments: High-Velocity Stars

Certain observations of peculiar stellar behavior caused definite problems with the Kapteyn model of our Galaxy. These peculiar stars—high-velocity stars—had been noted as early as the late nineteenth century, but the implication of their strange behavior in relation to the Kapteyn Universe was not recognized until the 1920s.

During the twentieth century, Benjamin Boss was the first to notice a highly asymmetric distribution in the directions of motion of high-velocity stars.[1] The asymmetry was confirmed by Walter S. Adams and Arnold Kohlschutter—astronomers famous for developing the method of spectroscopic parallaxes. As they noted in 1914, "A peculiar fact is the great preponderance of large negative over large positive velocities. ... No less than 75 per cent of the large velocities observed are negative."[2] Negative velocities indicate objects approaching the sun; positive velocities indicate objects receding from it. An asymmetry of such magnitude was certainly notable.

Adams and others investigated high-velocity stars further at Mount Wilson and in 1919 found additional evidence for the existence of the asymmetry. From improved parallax determinations, proper motions, and radial velocity measurements they were able to find the actual space velocities of 37 stars, which were selected because of their high radial velocities. Although the stars moved in many directions, they tended to avoid nearly an entire hemisphere between galactic longitudes $l^I = 131°$ and $322°$.[3]

Almost seven years later, a search of earlier literature revealed that the asymmetric distribution of the motions of high-velocity stars had actually been discovered as early as 1871 by a Swedish astronomer named H. Gyldén, who had noticed that proper motions tend to drift in one direction in one part of the sky. Furthermore, at right angles to the areas of greatest proper motions there is zero drift. Even though his data were not very accurate, Gyldén correctly attributed the effects to a rotation of our Galaxy.[4] Unfortunately, his work went unnoticed until after the effect was rediscovered and confirmed more precisely with improved data.

Jan H. Oort, who had studied under Kapteyn at Groningen in the Netherlands, naturally became interested in galactic systems; his early research dealt with the fascinating problem of high-velocity stars that appeared to be physically associated with the Milky Way. In 1922[5] he collected data on radial velocities and space velocities that confirmed earlier studies: between galactic longitudes $l^I = 162°$ and $310°$ all the high velocities were positive and in the opposite segment they were all negative.

A new discovery made by Oort was that the asymmetry is sharply dependent upon velocity. Below 62 km/sec, radial velocities have random directions, but above that critical velocity, the directions of motion exhibit a marked asymmetry. Oort tried to explain the critical velocity in terms of a dynamical theory involving a local system based on Kapteyn's model for our Galaxy. As Oort[6] admits, he did not find a satisfactory explanation because of his belief in a local system. Oort's doctoral thesis[7], presented in May 1926, was one of the last works on high-velocity stars before their connection to galactic dynamics was definitely appreciated. From calculations of the velocity necessary for a star to escape from a stellar sys-

Walter S. Adams. (Yerkes Observatory photograph)

tem like Kapteyn's universe, Oort concluded that the high-velocity stars had to be intruders into the system. Assuming that 62 km/sec was the escape velocity from a stellar system like that proposed by Kapteyn, Oort calculated that the average stellar mass had to be 0.65 solar masses instead of 1.0 solar mass, as Kapteyn had assumed; otherwise the velocity of escape would have been higher than 62 km/sec. Oort reasoned that since all the stars trapped by the gravitational system of our Galaxy would exhibit random motions, the boundary of symmetry at 62 km/sec indicated the limiting velocity at which stars were contained within the system. However, the same stellar system, assumed to be in dynamical equilibrium with an average radial velocity of 15 km/sec, required the average stellar mass to be 5 solar masses.[8] Oort noted this difference, but could offer no explanation for it.

In his thesis, he considered the possibility that our galactic system might be extended farther than the local group of stars. Previously, he had not referred to the topic in published papers. However, in his thesis, Oort listed the reasons[9] for believing in an extended system such as Shapley had proposed. Among them were the concentration of globular clusters toward the galactic plane and toward one direction in galactic longitude. He mentioned that other types of objects had the same concentration. He also noted that the globular clusters had the same systematic motion as the high-velocity stars.

Another point Oort discussed was also pertinent to observations of globular clusters. The velocities of 19 globular clusters were known when his thesis was written, and their average velocity was approximately 92 km/sec.[10] Oort noted that objects of such high velocities could not be gravitationally held by a system such as that proposed by Kapteyn. Clearly, then, they must belong to a larger system, some 200 times more massive than the Kapteyn Universe. This important point had long been overlooked by most astronomers. Oort also noted that the remaining postulated matter was not visible to astronomers, probably because it was hidden from view by obscuration in the galactic plane.[11] The question of obscuring matter was, of course, speculative, but it later became a prime issue in galactic astronomy.

An important advance in the concept of our Galaxy occurred when Oort reasoned that the systematic high velocities of the globular clusters indicated that we are moving rapidly with respect to the center of our galactic system, in a direction 100° from the direc-

Jan H. Oort: 1900–

Jan Oort, who began his studies at Leiden, the Netherlands, near where his family lived, went to study at Groningen, under the famous astronomer Kapteyn. This move was advised by an astronomer at Leiden who himself had been a student under Kapteyn. Oort's decision to leave Leiden had a lasting influence on his career. When he arrived in Groningen, he had not decided which field to choose—he was interested in physics as well as astronomy—but Kapteyn inspired him to begin scientific research in astronomy. Quite naturally, Oort entered Kapteyn's field of galactic astronomy, a move that eventually led to the discovery of galactic rotation and the determination of the size of the Milky Way.

Oort was also introduced at an early age to international cooperation, through Kapteyn's Plan of Selected Areas, which enlisted the help of many observatories in many countries. Oort also recognized the value of Kapteyn's close connections with Mount Wilson. Later in life, Oort ardently promoted international cooperation through organizations, as well as by encouraging the research of individual astronomers.

J. H. Oort (Yerkes Observatory photograph)

---

tion of maximum frequency of globular clusters.[12] According to his model, the high-velocity stars were intruders into our local system of stars, but were definitely associated with the larger galactic system, and were themselves rotating about the galactic center.

If we want to suppose the high velocity stars . . . to be members of the larger galactic system, we should have to assume that they are moving with respect to the center of this larger system, for they have a large systematic velocity with respect to the globular clusters and RR Lyrae variables. We might think of a rotation of the system of these high velocity stars. Our cloud would then move nearly in the direction of this rotation, with a somewhat higher velocity."[13]

Since Oort was not completely certain of the validity of his theory, even though it accounted for many observations, he added: ". . . I want to emphasize that the foregoing is not meant as an explanation but only as a possibly useful working hypothesis."[14]

The Lindblad-Oort Theory of Galactic Rotation

The study of high-velocity stars pointed to the conclusion that our Galaxy was rotating. Theoretical developments in galactic dynamics, however, were needed to incorporate the individual observations into

65

one scheme. The theorist who accomplished the task was Bertil Lindblad, of Sweden.

Lindblad proposed a rotating model for our Galaxy, which included the feature of subsystems rotating at different angular speeds. The result would be systems of stars revolving in elliptical orbits about the galactic center; the most elliptical orbits would be for stars with the lowest speeds at apogalaxion. Lindblad's model then explained observational details fairly well and now has proved to be essentially accurate.

Aside from the importance of his theoretical models and calculations, Lindblad had a direct influence on other astronomers. Oort, as a young researcher, was definitely influenced by Lindblad, as indicated in a letter written by Oort to Lindblad:

I must confess that I only got an insight into the real significance of your theory when reading some remarks in one of your latest papers on the subject. . . . The main objection I had to your theory was the same as that which I always felt against Kapteyn's and Jeans' theory of the galactic system, namely that it implied systems penetrating each other and rotating with respect to each other. It was considerably later that I saw that the different velocities of rotation as required by *your* hypothesis could arise in a quite natural way. I then felt at once convinced that there *had* to be some truth in your suggestion because of the un-artificial explanation which it gave of the enigma of asymmetry."[15] (Oorts' emphasis.)

Although Lindblad had stated his theory before, Oort had not understood its implications for a very simple reason—Lindblad's mathematical treatment was extremely obscure, like that of some theoreticians who are sometimes less concerned with the real world than with an abstract one. Oort, who had been thoroughly trained in physics, astronomy, and mathematics, was nevertheless an observer with an intuitive feeling for dynamics.

After he had accepted Lindblad's concept of rotating subsystems, Oort calculated the consequences of a galaxy rotating in a different manner.[16] As he states in his article, "it is easily seen that" the radial velocity V_r can be expressed by

$$V_r = rA \sin 2 \, (l - l_0),$$

where A is a constant, r is the distance of a star, and $(l - l_0)$ its angular distance from the galactic center at galactic longitude l_0. A student under Oort at the time remembers a slightly different version of the derivation of the equation. As the story goes, Oort presented the dynamics problem to the class and they all worked for several days before arriving at a simple expression. Here, as in many other scientific papers, "it is easily seen" should be prefaced by "once the solution had been found."

A similar expression, with an additional constant B, applies to proper motions, and from the two equations the parameters associated with the galactic dynamics can be determined. In particular, the distance to the galactic center R_0 can be calculated. Originally, Oort found a value for R_0 of about 5.1 to 5.9 kpc. The direction of the center he found to be

$$l_0^{\text{I}} = 323 \pm 2.4 \ ^{17}$$

Later, in September 1927, Oort again determined the distance R_0 using different data; the results were similar:

$$R_0 = 6.3 \pm 2.0 \text{ kpc } ^{18}$$

The similarities between Oort's determinations and those of Shapley are striking—they both place the galactic center within a few degrees of each other, and they both have the same order of magniture for the size of the Galaxy. The differences, however, should also be noted. Oort's value for the distance to the galactic center is one-third Shapley's value. Although such factors of two or three often were not significant in the early stages of astronomical theories because of the difficulties in obtaining accurate data, in this case the difference was significant, being caused by a fundamental omission from the theories—namely, the effects of absorption had been ignored.

Just as Lindblad's theory inspired Oort to work out the details of differential galactic rotation and to find observational confirmation, Oort's published articles inspired other astronomers to pursue the same area of research. Notably, J. S. Plaskett at the Dominion Astrophysical Observatory in Victoria, British Columbia, was inspired to apply the data from his studies of hot, O and B type stars to further confirm the Lindblad-Oort galactic rotation theory. As Plaskett wrote to Oort shortly after the publication of the original confirmation: "Your recent splendid and most interesting work on the rotation of the galaxy has induced me to apply your analysis to the radial velocities of some of the faint B stars recently observed here. . . ."[19] O and B stars are well suited for determination of Oort's constant A. As mentioned before, the relative accuracy for determinations from radial velocities increases with distance. Since O and B stars are intrinsically extremely bright, they can be detected at great distances.

J. A. Pearce and Plaskett[20] had been studying O and B stars before Oort's announcement on galactic rota-

The 72-inch reflector at the Dominion Astrophysical Observatory at Victoria, British Columbia, Canada. (H.C. King, *History of the Telescope*, Sky Publishing Co.: Cambridge, Mass., 1955)

J.S. Plaskett. (Yerkes
Observatory photograph)

tion and had obtained radial velocities of many different stars. Immediately, Plaskett was able to apply Oort's analysis and determine the direction of the galactic center and the value of A. The results he obtained were close to those of Oort. For l_0, Plaskett found a value of $324° \pm 1.8°$, which is only $1°$ from Oort's value. For the constant A, Plaskett calculated a value of 15.5 ± 0.7 km/sec/kpc; Oort had determined A to be 19 ± 3 km/sec/kpc.

Oort was pleasantly surprised with Plaskett's paper.

I am glad to see that Lindblad's suggestions on the explanation of the motions of the high velocity stars, by which suggestions so many facts not hitherto understand [sic] are so beautifully linked together, are being appreciated elsewhere.

. . . I must say that I am almost surprised at the accuracy with which your rich and homogeneous material of faint B- and O-stars confirm the rotation effect. I had not expected that so much important material would so soon be available.[21]

The vast amount of spectroscopic data on O and B stars that Plaskett and Pearce had collected confirmed the analysis that Oort had applied to relatively few stars.

68

Although the Lindblad-Oort theory of galactic rotation and the subsequent confirmation of that theory by Plaskett and Pearce indicated that Shapley's model might be too large and that Kapteyn's model might not be correct, the discrepancy of scale still remained. In the years between 1920 and Oort's presentation of his analysis of galactic dynamics, Shapley had worked to reduce the uncertainties in his distance estimations. Consequently, many astronomers were ready to endorse the Shapley model. Nevertheless, the results from the massive collection of data from Victoria that complemented Oort's analysis presented an unresolved conflict. Once again, determining the accuracy of the various techniques used to estimate distance was of prime importance.

John S. Plaskett: 1865–1941

J. S. Plaskett's career in astronomy began late in life—he was about 40 years old—but his previous training served him well in the role he was to play. His early training had been in mechanics and for years he had worked at the Toronto University as mechanical assistant in the physical laboratory. In the year 1903, when he began to work in Ottawa at a newly established observatory, he went not as an astronomer but as a mechanical superintendent. Within a few years, however, his research capacities were recognized because of his work on stellar spectra. By 1910 he had been appointed to three committees of the International Union for Cooperation in Solar Research; in 1913 he was entrusted with the construction of a large reflecting telescope for the Canadian government.

His mechanical training was invaluable during the planning of the telescope and by 1918 a 72-inch reflector was in operation at Victoria, British Columbia. Designed for spectroscopic research, the instrument performed extremely well, and produced a tremendous number of spectrograms each year.

Plaskett was not a great theoretician, but he was an efficient, patient, and careful observer. He did not seek out the glamorous research fields, but instead was content to do the essential routine cataloguing that proved to be extremely valuable during the 1920s. Not only did Plaskett's spectrograms provide the impetus toward solving the mysteries of interstellar calcium, they also provided the basis for rapid confirmation of Oort's theory of galactic revolution.

Dark and light nebulosites in the constellation Monoceros. (Hale Observatories photograph)

CHAPTER 3

Absorption and the Solution of
the Galactic Scale Problem

Early Evidence of Absorption

As we noted in Chapter 1, Kapteyn realized that his model of our Galaxy was valid only if no general absorption of light occurred in space. If absorption in the amount of 1 to 2 magnitudes per kiloparsec were present, then the apparent decrease in the density of the stellar distribution with distance from Sun could be explained without postulating a real physical decrease. Furthermore, the suspect central position of the Sun would be readily explained. Consequently, Kapteyn was especially interested in determinations of the amount of absorption. Early evidence of the existence of interstellar absorbing material was provided in an extensive photographic survey of the Milky Way.

In 1889, Edward Emerson Barnard, using first a 6-inch and then a 10-inch camera lens, began to photograph star clouds in the Milky Way. These star clouds were actually areas along the Milky Way in which the stars are so closely packed that it is hard to distinguish one star from its neighbor. Dark areas which appear in these star clouds had been known to exist as far back as the time of William Herschel who is credited with having called these areas "holes in the heavens," as though they were devoid of stars. What the holes actually were was not certain in 1889 when Barnard began his study; they could have been regions sparsely populated with stars or areas where something was blocking the stars from view.

Barnard became interested in these dark areas and for many years studied them in his photographs of Milky Way star clouds. One of the most convincing pieces of evidence that obscuring masses do exist was the relation noted by Barnard in 1905 between dark areas and visible nebulae. "For many years I have called attention repeatedly to the fact that many of the nebulae occupy vacant regions as if their existence was in some way the cause of the scarcity of stars."[1]

Even so, Barnard was not yet convinced, as his remarks in the same paper indicate: "Though this may in a few cases be true . . . I think they can be more readily explained on the assumption that they are real vacancies. . . . In the few cases where the appearance would rather suggest the other idea . . . the evidence is still not very strong."[2]

By 1907 Barnard realized his error, after studying a remarkable photograph showing nebulosity connected to a very dark lane located in the constellation Taurus. At that time, however, he did not accept the theory that all the dark markings were due to obscuring bodies: "I have been slow in accepting the idea of an obscuring body to account for these vacancies; yet this particular case almost forces the idea upon one as a fact. . . . There is no question that this is real, and not a subjective effect."[3]

Although, as photographic evidence accumulated, Barnard became more receptive to the existence of obscuring matter, he still found it difficult to accept the vast scale for matter implied by his photographs. (He did realize that if the matter existed, it was intimately connected with diffuse nebulae.)[4] By 1919,

E.E. Barnard *c.* 1885. (Yerkes Observatory photograph)

71

however, Barnard had reversed his opinion: "I do not think it necessary to urge the fact that there are obscuring masses of matter in space. This had been quite definitely proved by my former papers on this subject."[5]

A majority of astronomers now agreed that large clouds of obscuring matter did exist—diffuse matter is found in large quantities in space. The distribution of the matter, however, became a question of debate —was all diffuse matter collected in clouds, or was there a component spread thinly throughout all galactic space? The answer to that question had to be found before results on the structure of our galactic system became definite.

Realizing that a direct determination of general absorption would be difficult, Kapteyn proposed other methods, the most important of which was to search for reddening of light due to wavelength-dependent scattering. Kapteyn stated in 1909 that "there can be no reasonable doubt but that the violet end of the spectrum must be more strongly affected than the

Edward Emerson Barnard: 1857–1923

E.E. Barnard at the 36-inch refractor at Lick Observatory, c. 1900. (Lick Observatory photograph)

Born in Nashville, Tennessee, after his father's death, Barnard suffered extreme poverty and hardship in his youth. During the Civil War he and his family survived on army hardtack found floating down the river. Since he was forced to start work at an early age, he had little formal education. He found employment at the age of nine in a photographic studio in Nashville. The experience he gained there served him well in later years.

His interest in astronomy began in 1876, when he read a stray copy of Dr. Thomas Dick's *Practical Astronomer*. The next year he purchased a small telescope and began studying the sky. In 1881, he discovered a comet, and found several more in the following years.

Although he lacked formal education, Barnard had studied on his own, and thereby won a fellowship in astronomy at Vanderbilt University. There he studied languages, mathematics, and the natural sciences. He also continued his astronomical studies at a small observatory, and his fame as an astronomer grew. As a result of his research, Barnard was chosen to be one of the original astronomers at the newly founded Lick Observatory at Mount Hamilton, California.

Barnard did not limit himself to observing only one class of objects: he studied comets, double stars, planets, and nebulae. Perhaps his most famous achievement is his collection of photographs of the Milky Way star-clouds. Through his study of the dark areas that appeared in the star-clouds, Barnard recognized that large amounts of obscuring matter exist in space.

Astronomers involved in galactic research (l. to r.): A.S. Eddington, J.S. Plaskett, W.S. Adams, J.H. Oort, H.N. Russell, Miller (Director of the Plymouth Chamber of Commerce), F. Dyson, F. Slocum, and B. Lindblad. The photograph was taken during an International Astronomical Union outing to Plymouth, Mass. in 1942 (Sky Publishing Co. photograph)

less refrangible rays."[6] It seems probable that he was considering the effects of Rayleigh scattering, which is proportional to λ^{-4}. Since this function was the best understood at the time, it was a likely candidate.

Later in 1909, Kapteyn compared the magnitudes of stars obtained visually and photographically.[7] The photographic plate is more sensitive than the eye to blue light, so that the difference between the two magnitudes is an indication of a star's color. He found a small reddening in stars with increasing distance. Analysis of his results, on the assumption that they were due to Rayleigh scattering, revealed absorption of about 0^m3 kpc in the visual magnitude,[8] but this value depended strongly on the validity of his assumption.

For several years Kapteyn worked on the problem but was plagued by bad weather and ill health.[9] However, his advice to search for reddening in stellar spectra had not gone unheeded; other astronomers had taken up the task and collected a considerable amount of data. By 1914, he was ready to review the studies of other astronomers and make a statement about the existence of absorption. Also, the time was right for such an article, as George E. Hale noted in a letter to Kapteyn: "With the new material now available, I hope you will be in a position to complete your paper on the general question of space absorption, as the 'psychological moment' for its publication seems to have arrived."[10] Hale realized that astronomers were ready to accept the concept of interstellar

Nebulous region in Taurus—obscuring matter blocks out the background stars. (E.E. Barnard photograph, 1907)

absorption as important to astronomical calculations, although previously the prevailing attitude had been one of indifference.

After reviewing the data available,[11] Kapteyn summarized the results as follows:

1. On the average the apparently fainter stars are redder than the bright ones.
2. Apparent magnitude and spectral lines being the same, the stars are redder the farther away they are.[12]

If apparently fainter stars are redder than the bright ones, then there are three possible explanations:

1. Late spectral type stars, i.e., red stars, are predominantly among the fainter stars.
2. The absolute brightness of a star could affect its color index, i.e., the absolute luminosity of a star could possibly affect the amounts of light given out in the various portions of the spectra, even though the strengths of characteristic spectral lines remained the same as in less luminous stars.
3. Absorption selective with wavelength could redden a star's spectrum, i.e., bluer wavelengths could be scattered to a greater extent than red wavelengths.

If more distant stars are redder, when one considers only stars with identical apparent magnitude and spectral line features, then the last two explanations are the only ones possible.

Kapteyn concluded that the evidence indicated that the two observational effects were real and that the relative frequencies of the spectral classes were the key to distinguishing between explanations. Since the relative frequencies were not known conclusively, Kapteyn would not commit himself definitely. He did

74

imply that he believed in absorption, however, by mentioning other investigations indicating that matter existed in space and was scattering light. The important question to be answered was, of course, how much absorption existed?

As we have stated before, the main reason for Kapteyn's interest in absorption was his desire to study the structure of our galactic system. His first estimates of the amount of absorption were directed at explaining the apparent location of the Sun at an exceptional position in space. Kapteyn at that time (1904) was unwilling to believe that we should occupy such a privileged position and he evoked absorption amounting to 1^m6/kpc to rationalize his belief. His later investigation, in 1909, however, yielded an absorption coefficient of only 0^m3/kpc. His studies after that did not provide the increase in the coefficient necessary for compatibility with a universe of constant star-density. Eventually, around 1915, Kapteyn began to change his opinion of the galactic structure.

In a letter to Hale, written in 1915, Kapteyn admitted that he was forced to believe in the reality of the apparent decrease in star density with distance from the Sun.

One of the startling consequences is, that we have to admit that our solar system must be in or near the center of the universe, or at least to some local center.

Twenty years ago this would have made me very sceptical. . . . Now it is not so—Seeliger, Schwarzschild, Eddington, and myself have found that the number of stars

Filamentary nebula in Cygnus. (Hale Observatories photograph)

is greater near the sun. I have sometimes felt uneasy in my mind about this result, because in the derivation the consideration of scattering of light in space has been neglected. Still it appears more and more that the scattering must be too small and somewhat different in character from what would explain the change in apparent density. The change is therefore pretty surely real.[13]

Affairs took a sharp turn after Harlow Shapley published the results of globular cluster studies indicating that selective absorption, i.e., reddening, was less than 0^m1/kpc.[14] For many astronomers, the question of absorption became a null issue; e.g., Shapley listed Kapteyn, Hale, and Hertzsprung—three notables of the time—as members of this group.[15] The zero absorption derived from Shapley's studies seemed to give the death blow to the concept of interstellar absorption. Kapteyn, whose research had failed to detect any significant amount of absorption, accepted Shapley's evidence as conclusive.

Ironically, Shapley's determination of zero absorption reinforced Kapteyn's arguments for the "Kapteyn Universe." Earlier, he had evoked absorption to eliminate the uncomfortable heliocentric aspect of his galactic model; now he had no choice but to interpret his star counts in terms of a Galaxy ten times smaller than envisioned by Shapley.

In spite of the general acceptance of Shapley's absorption result, acceptance of the transparency of space met some difficulties. The major difficulty was the lack of globular clusters and spiral nebulae in the region of the galactic equator. Either such objects were indeed absent from that region or they were hidden from view. To explain the observations without recourse to absorption, Shapley had to propose speculative disruptive forces in the galactic plane.[16] For his speculative efforts, Shapley was criticized. The influential George Ellery Hale cautioned Shapley on the dangers of making daring hypotheses without presenting supporting proof.[17] And Henry Norris Russell—Shapley's friend and former professor—evoked the admonishment *"Entia non multiplicanda praeter necessitatem"*[18] (Entities must not be multiplied beyond necessity).

Other astronomers, notably Heber D. Curtis (who played an important role in the account in Section One), resorted to obscuring matter to correlate the region of avoidance with dark equatorial bands commonly found on photographs of edge-on spiral nebulae. Reasoning by analogy with the dark equatorial bands, Curtis concluded that occulting matter existed in rings or whorls in the outer regions of our own Galaxy. This obscuring matter, he reasoned, prevents our seeing the equatorial global clusters.[19]

Dark clouds, such as those photographed by Barnard, were suggested as being responsible for the region of avoidance. Russell,[20] in particular, liked the suggestion and he displayed remarkable insight into the problem of interstellar matter and absorption. In a letter to Shapley, he correctly guessed several properties of interstellar matter:

. . . dust or fog is a far more powerful absorber per unit mass than gas, and most of the matter in interstellar space ought to be solid—if its chemical composition is in the least like that of the stars—but may be finely divided.[21]

Shapley did not agree with Russell. The distribution of globular clusters in space, argued Shapley, indicated that the region of avoidance was a slab of approximately constant thickness.[22] If obscuring matter were responsible for the lack of equatorial globular clusters, the region of avoidance should be wedge-shaped. To support his argument Shapley produced a diagram of the positions of globular clusters, which illustrated the geometric problems involved. Russell was not convinced, however, and wrote that "the diagram . . . is not competent evidence either for or against the reality of the zone of avoidance . . . you have jumbled together the different galactic latitudes, without distinction."[23]

To satisfy Russell's criticisms, Shapley sent another diagram outlining the positions of the globular clusters in a 70° segment of galactic longitude and he incidentally included the positions of Cepheid variables and open clusters in relation to the region of avoidance. The globular clusters in the diagram extend to a distance of more than 50 kpc; the open clusters extend only to about 17 kpc. Significantly, Shapley wrote in the margin of the diagram, "Why do we not find many open clusters further from the sun???"[24]

Russell immediately recognized the importance of Shapley's diagram as a demonstration of obscuring matter and in reply he remarked that "one can almost see the absorbing matter in the galactic plane, cutting off the remoter open clusters from view."[25] Although there were still many questions unanswered about the region of avoidance, there was enough evidence to convince some astronomers that interstellar absorption could not be neglected.

For many years after Shapley's announcement of zero interstellar absorption, based on studies of globular clusters, research in the field of absorption was sporadic and uneventful. The astronomical community remained split in its views on absorption. Those who believed Shapley's interpretation naturally accepted the concept of a transparent space. Other astronomers still had doubts that a transparent space was compatible with observations such as the region of avoidance.

Indirect Evidence of Absorption

Obscuration, and the reddening of light by scattering from particles, were not the only effects sought to confirm the possible existence of absorption. Detection of interstellar gas was considered a possible route by which general absorption could be theoretically calculated, since the scattering process for gaseous material, i.e., Rayleigh scattering, was well known. In 1908, Kapteyn considered this approach.

Owing to the gas of the corona lost by the sun, to similar loss presumably suffered by other stars, to that lost by comets, etc., interstellar space must contain, at every moment, a considerable amount of gas. Might not this gas, in a thickness of hundreds of light years, cause an appreciable absorption of light? . . .

Even more important than the general absorption . . . would be a gas-absorption, producing *space-lines*. . . . If there are spacelines, they must not share in that part of the radial motion which is due to the motion of the stars themselves. . . .

As, however, I have no evidence as to the real occurrence of such lines or bands, no more need be said about them at present.[26] (Kapteyn's emphasis.)

Actually, although Kapteyn did not know it, space lines had already been discovered. In 1904, the astronomer Johannes Hartmann observed that in the rapidly revolving binary star δ Orionis, the strong lines of ionized calcium did not participate in the Doppler-shift oscillations of the hydrogen and helium lines. If the calcium lines had arisen in the atmospheres of the two stars comprising δ Orionis, they should have oscillated as the stars orbited each other. Hartmann also noticed that the calcium lines were distinctly narrower than the other lines; he named these non-oscillating lines "stationary lines."

Shortly after Kapteyn made his predictions about space lines, Vesto M. Slipher found that spectrograms of the binary star β Scorpii contained sharp calcium K lines, whereas its other lines were broad. Furthermore, the K lines did not participate in the oscillations of the other lines. Since Slipher remembered Hartmann's similar case, he extended the investigation to

A gathering of astronomers (foreground, l. to r.): Dr. McBride, J.C. Kapteyn, K. Schwartzschild, V.M. Slipher. (Yerkes Observatory photograph)

other stars. Soon Slipher had found several cases of binary-star spectra containing narrow "stationary" calcium lines. For all of these spectra, the absolute velocities of the calcium determined from Doppler shifts, corrected for solar motion and the Earth's orbital motion, indicated that it was practically stationary with respect to the local system of stars.

From the data he collected, Slipher concluded that the calcium thus observed was definitely outside the solar system and apparently in interstellar space. He suggested that clouds of calcium gas were interposed between the stars and the earth: "May we not be observing in these calcium lines the 'space lines' which Kapteyn's researches have led him to predict, and the phenomenon be due to selective absorption of light in space?"[27] Kapteyn, of course, was pleased with Slipher's results and sent him warm congratulations.[28]

Slipher's investigations and interpretations were substantially correct. Unfortunately, they did not receive the attention they deserved, and quickly slid into obscurity.

In the year 1920, R. K. Young—an astronomer at Victoria, British Columbia—noticed that the stationary lines were found only in the spectra of hot stars, usually of class B3 or earlier. [The term "earlier" is a relic of a time when the main sequence was believed to be an evolutionary track starting with O and B stars and descending through A, F, G, to K and M stars.] He also noted that the phenomenon seemed to

J.S. Plaskett supervising the resilvering of the 72-inch mirror at Victoria. (*Popular Astronomy 28*, 1919)

The deep snows of 1918. An astronomer on his way to the Yerkes Observatory for a night's work. (*Popular Astronomy, 26*, 1918)

be associated only with spectroscopic binaries, i.e., close, rapidly revolving binary stars detected by characteristic shifts in the spectral lines. Young concluded that the lines were due to ionized gas clouds that surrounded the binary stars; differences between the space and stellar lines Young ascribed to errors of measurement.

Young's assumptions and conclusions were short-lived; in the same year, and also at Victoria, Plaskett found contradictory evidence. While investigating a massive binary star system, he[29] found that the stationary lines indicated a velocity that differed from the average motion of the pair by nearly 40 km/sec. There could be no possibility that so large a difference could be attributed to measuring errors. In addition, Plaskett found that binaries were not the only stars to have the characteristically narrow calcium lines: in general, stars earlier than B3 had the lines, often displaced by as much as 40 km/sec.[30] There seemed to be every indication that clouds of calcium gas existed everywhere in space and that the O and B stars were rapidly moving through the clouds. Furthermore, the calcium clouds appeared to be nearly stationary, as Slipher's results had previously indicated. As a consequence, Plaskett interpreted the data as indicating an interstellar origin for the lines. He suggested that the calcium was ionized as it was approached by hot O and B stars, and hence gave rise to the Ca II absorption lines.

Some confusion was contributed to the problem by a series of papers by Otto Struve at Yerkes Observatory.[31] He carefully analyzed available homogeneous data and formulated logical conclusions; there were spectroscopic results, however, that led him astray. Previous investigators had noticed that the interstellar calcium lines associated with some spectroscopic binary stars apparently oscillated, although the amplitudes were smaller than those of the stars. Plaskett had suggested that the results might be spurious, the result of difficulties in separating the interstellar lines from those originating in the stellar atmospheres. Struve did not agree and, furthermore, he postulated that the revolving binary stars were creating eddies in the interstellar gas cloud.

As interest in the interstellar medium mounted in the mid-1920s, the theorist Arthur S. Eddington focused his attention on the problems associated with the interstellar absorption lines. He carefully considered the data available, made a few assumptions, and produced a theory which could be used to make predictions about the behavior of the lines.

For the Bakerian Lecture before the Royal Society in 1926, Eddington discussed theoretically the physical conditions of diffuse matter in interstellar space. He noted that although most astronomical studies neglected the effects of matter in space, definite evidence for its existence was present. First, he called attention to nebulae, which undoubtedly contained diffuse matter; second, he stated that the phenomenon of fixed calcium and sodium lines in certain stars is probably due to absorption by a diffuse cloud in space.[32]

In his lecture, Eddington referred to the investigation of Plaskett, which proved that the calcium cloud was moving rapidly with respect to individual stars, yet had a low velocity with respect to the average motion of the local stars. Eddington agreed with Plaskett that the calcium clouds were not moving with the stars, but disagreed with other parts of Plaskett's interpretations. Plaskett had hypothesized that large clouds of calcium must exist in space through which

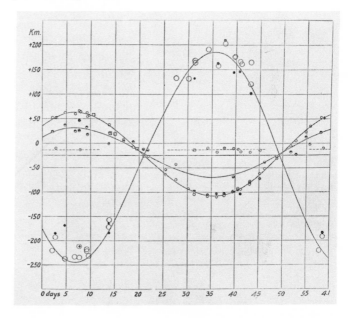

the stars passed. Eddington concluded, on the other hand, that the calcium lay in a continuous cloud that pervaded the entire galactic system and was nearly at rest relative to the average stellar motion.

Eddington used physical concepts to great advantage in attacking astronomical problems. From a calculation of the energy density of stellar radiation in interstellar space, he was able to estimate the temperature of the diffuse gas and also the relative proportions of the various states of ionization in calcium. He concluded that the temperature of the cal-

The radial-velocity curves of a binary star. The large and small open circles represent the secondary and primary stellar velocities respectively. The filled circles represent stellar calcium data; the dotted circles represent interstellar calcium data. (*Publications of the Dominion Astrophysical Observatory, 5*, 1933)

cium gas was high enough for most of the calcium to be doubly ionized.

It is interesting to note that Eddington's theory spelled the doom of the theories demanding that the radiation from the hot O and B stars singly-ionize the calcium atoms. Close to the stars, the gas is so hot that nearly all atoms of calcium are doubly-ionized—there are so few singly-ionized atoms that they could not contribute much towards the observed line absorption.

Another triumph of Eddington's theory was its explanation of the observation that interstellar absorption lines were limited to B3 or early-type stars. He was able to provide three reasons for their absence in late-type stars:[33] first, the calcium lines arising in the stellar atmosphere are broad in later stars and hence mask the interstellar lines; second, the late-type stars observed do not have large radial velocities (without large radial velocities, the stellar and interstellar absorption lines will not be separated); third, the late-type stars are intrinsically fainter and hence cannot be seen at great distances. (Unless the stars are at great distances, there are not enough singly-ionized calcium atoms in the intervening space.) The combination of these three reasons does not make it impossible to detect interstellar lines—any distant star with a high enough velocity at a great distance will show the lines—but with the instrumental limitations of spectroscopy their detection would be extremely difficult.

Besides providing the first firm theoretical explanation of interstellar absorption lines, Eddington's theory stimulated more research. The data hitherto collected were not satisfactory for any detailed examination of the interstellar medium; more complete studies had to be undertaken.

Intensity of the interstellar calcium K line as a function of apparent magnitude. (*Astrophysical Journal, 65,* 1927, University of Chicago Press)

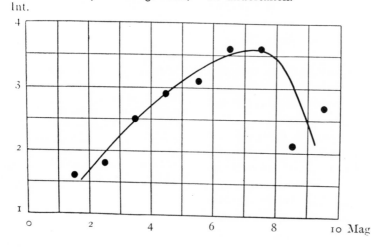

The Yerkes 40-inch telescope displayed at the Chicago World's Fair of 1893. (Yerkes Observatory photograph).

Otto Struve, who had previously been engaged in research at Yerkes Observatory on the interstellar calcium lines, published a series of papers on the subject in the years following the Bakerian Lecture. In 1927, Struve[34] found that certain regions of the sky seemed to be more favorable than others to the production of the interstellar lines. He interpreted this as favoring the hypothesis that individual clouds of calcium atoms, rather than a continuum, existed throughout space, since in some directions there were more clouds than in others, giving rise to the greater intensities of the absorption lines.

Struve also noticed that the intensity of the calcium line increased with increasing stellar magnitude, until about seventh magnitude. After that, the intensity appeared to decrease again. Since all stars under investigation had about the same absolute magnitude, the data indicated that the line intensity increased with

Alvan Clark and Carl Lundin with the 40-inch lens. (*Astrophysical Journal, 6,* 1897, University of Chicago Press)

distance until about 600 parsecs and then dropped off again. Struve commented that "the decrease in intensity for distances greater than 600 parsecs, if real, would be contrary to Eddington's hypothesis. It will be interesting to obtain a larger amount of material for stars fainter than the seventh magnitude, which would settle this point definitely."[35]

Struve did not accept Eddington's theory, for reasons other than that mentioned above. He found that the calcium lines disappeared abruptly at type B3 stars; the intensity of the lines started to fall off at spectral type B2, and rapidly decreased to near zero for B3 and later. Struve indicated that he believed Eddington's theory necessitated a gradual drop in intensity; hence the sharp disappearance seemed to be fatal to the theory.[36] Struve preferred Plaskett's original theory and the assumption that absorption occurs only in the vicinity of hot stars. From such an assumption, Struve concluded that the density of calcium clouds increased up to a distance of 600 parsecs and then decreased again.

Struve was mistaken mainly because his data were not complete enough to warrant any definite conclusions, a fact that he recognized in the end of his paper.

Present observational material does not make it possible to accept definitely either hypothesis. It is therefore advisable to accept the effects due to distance, spectral type, region of the sky, etc., as properties of the calcium clouds without, at present, attempting to specify the physical explanation.[37]

The following year Struve published the results of more extensive data in which several of his objections to Eddington's theory were eliminated. Significantly, Struve found that the decrease of intensity he had reported beyond 600 parsecs was not real, and the new data indicated that the intensity behaved as Eddington had predicted: it increased monotonically. Struve was forced to admit that "our final conclusion would . . . be that present observational evidence does not contradict Eddington's theoretical ideas, but in some respects even favors them."[38] He still had some objections to Eddington's hypothesis, however, and referred to his previous objection to the sharp decline in line intensity at B3 type stars.

During his investigations, Struve had noted another phenomenon; his data revealed interesting residual radial velocities of the calcium lines after correction for the solar motion. In a footnote he considered the possibility "that the rotation of the galaxy advocated by

J. H. Oort . . . and recently confirmed by J. S. Plaskett . . . would satisfactorily explain at least part of these peculiar velocities."[39]

Perhaps an interest in the interstellar absorption lines as a means of confirming his own theory led Oort to take an active part in the development of the study. Shortly after Struve's 1926 paper was published, Oort wrote Struve[40] and requested further information on his studies, as well as permission to use the intensity data relating to proper motions to check the distance-intensity relation of the interstellar calcium absorption lines. Struve replied by sending the data requested and a comment on the state of his research.

> You may be interested that Dr. Gerasimovic and I have subjected the Ca problem to a more theoretical analysis. The results come out in favor of the idea of an interstellar cloud in the sense of Eddington. I have also analyzed all the evidence available for Na. Finally, we have made a new computation for your rotational effect, that the ampli-

---

Otto Struve: 1897–1963

It is not surprising that Otto Struve became an astronomer. His father was the professor of astronomy at Kharkov, Russia; one uncle was director of the Königsberg Observatory in Prussia; another uncle was director of the Berlin-Babelsberg Observatory; and his great-grandfather was one of the most famous astronomers in Russia. Otto's career in Russia, however, was interrupted by World War I. He served in the Imperial Army and the White Russian Army, and after their collapse in 1920 was forced to flee to Constantinople. Like many of the Russian refugees of the time, he suffered many hardships. Struve was lucky, however, in that his name was known throughout astronomical circles, and the director of the Yerkes Observatory provided him with an assistantship so that he could come to the United States.

At Yerkes, Struve received his Ph.D. and completed much of his later research. It was there that he investigated many aspects of the interstellar calcium clouds.

It was obvious from the start that Struve's great passion was astronomy: he worked at a tremendous pace and had little patience with persons of lesser enthusiasm. In his fervor, he was able to contribute to many different fields of astronomy: however, he was rarely satisfied to concentrate on the mundane task of working out the less important, but nevertheless necessary, details. This habit, as well as his tendency to publish speculative theories—some quite wild—to explain his observations, more than once irritated other astronomers. But no one can deny that his imaginative approaches to research served astronomy well.

Stages in the construction of Yerkes Observatory: *above:* 4 April 1896, awaiting completion of the domes. G.E. Hale's dog, Sirius, sits in the foreground; *below:* Construction of the dome for the 40-inch telescope; *opposite:* Installation of the mounting for the 40-inch telescope. (Yerkes Observatory Photograph)

tude of the oscillation in the Ca radial velocities is very nearly equal to one-half of that for the stars. This cannot be due to blending, since the effect is largest in the O stars, where there should be no blending at all.[41]

The most significant result of their joint research was the finding that interstellar absorption lines participated in the galactic rotational effects. This result reconfirmed Oort's differential rotation theory and, furthermore, provided evidence that interstellar gas existed uniformly throughout the Galaxy. Gerasimovic and Struve determined that the value of $\bar{r}A$ for the gas is one-half the value for the stars. (For the gas, they determined $\bar{r}A$ to be 5.3 ± 1.0 km/sec; for the stars, a value of 12.0 km/sec—almost exactly twice the former value.[42]) If A were constant, then, on the average, the gas would be one-half the distance of the stars, as would be expected for a continuous gaseous medium.

Shortly after Gerasimovic and Struve's joint publication appeared, Struve wrote Oort that additional confirmation of their results had been announced by J. A. Pearce and J. S. Plaskett at the Ottawa meeting of the American Astronomical Society.

. . . they are now convinced of the correctness of the idea that interstellar Ca$^+$ is everywhere in space. They confirm my results quite definitely, and they use the method of galactic rotation for it. Your factor rA comes out exactly 2.00 times greater for the stars than for the Ca radial velocities. . . . The precision of the work is wonderful.[43]

Plaskett and Pearce actually had data, before Struve and Gerasimovic's publication, indicating that the calcium participated in the galactic rotation as derived

86

by Oort. In fact Struve, who was a frequent visitor at Plaskett's observatory, may have learned about the rotation effect from Plaskett himself. Plaskett never forgave Struve for taking credit for the discovery.

It should be remembered from the last chapter that Plaskett, who had been studying O and B stars, had been able to provide almost immediate confirmation of the rotational effect predicted by Oort for radial velocities. Plaskett and Pearce were also able to utilize the O and B data to confirm the ratio of two (their mean value for the ratio was 2.01!) between Oort's $\bar{r}A$ factor for stars and for interstellar gas:

. . . comparing the rA for stars and clouds . . . the remarkable relation is again found that its value for the stars is almost exactly double that for the clouds. . . . When we consider that this relation holds for star groups of distances varying between 600 and 1600 parsecs, there can be no possible doubt that *Eddington's hypothesis of uniform distribution of the interstellar matter is fully confirmed.*[44] (Emphasis in original.)

Although the interstellar distribution of the gas, originally discovered by Slipher, had been clearly demonstrated by the year 1929, calculations of the amount of interstellar gas showed that it could not be a significant source of general absorption. Consequently, the discrepancies between Kapteyn's, Oort's and Shapley's galactic scales remained unresolved; the size of our Galaxy remained uncertain.

Confirmation of General Absorption

Although no conclusive evidence of general absorption was found during the 1920s, and although Shapley's research indicated negligible absorption, some astronomers were certain that absorption must exist. Oort[45] recently has stated that the region of avoidance convinced him of the existence of interstellar

Relation between the intensity of the interstellar calcium K line and Oort's rotational term $\bar{r}A$ (*Publications of the Dominion Astrophysical Observatory, 5*, 1933)

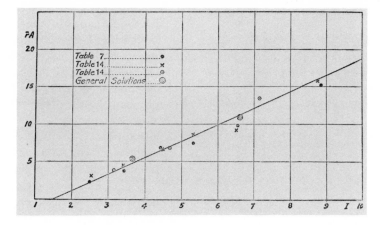

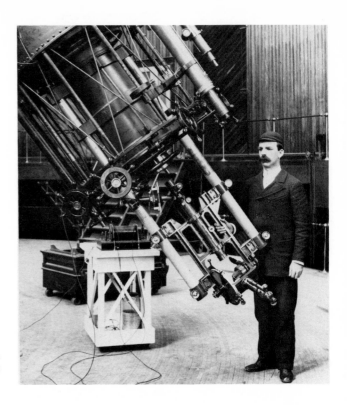

W.W. Campbell with the 36-inch refractor at Lick Observatory (Lick Observatory photograph)

absorption, a statement that is borne out by his publications. In 1927, he commented on the contradiction between the distance to the galactic center he had deduced, and the distance implied by Kapteyn's model of our Galaxy:

The most probable explanation is that the decrease of density in the galactic plane indicated for larger distances is mainly due to obscuration by dark matter. Such a hypothesis receives considerable support from the marked avoidance of the galactic plane by the globular clusters, a phenomenon for which up to the present time no other well defensible explanation has been put forward.[46]

Oort did not mention the effects absorption would have on Shapley's model.

Charles D. Shane was one of the first astronomers to realize and discuss the effects absorption would have on Shapley's model. About 1928, he taught that absorption would considerably shrink the distances being proposed by Shapley.[47] The attitude of many astronomers was beginning to change toward the reality of absorption. Word was also being circulated that the preliminary results of a study by Robert Trumpler indicated that absorption definitely existed.

Opposition to the concept of general absorption did not abate. For instance, Shapley did not relinquish his belief in the transparency of space. In 1929, although he acknowledged the fact that completely opaque bodies existed in space, concentrated along the galactic equatorial plane, he further suggested that the areas of complete obscuration could easily be detected and had well-defined boundaries. Outside of these areas, Shapley stated, space must be transparent, and he suggested a means of testing for transparency.

> Fortunately we can test the transparency of the Galaxy in any specific direction with the aid of extragalactic nebulae . . . in the places where outside galaxies are seen the apparent magnitudes are not seriously dimmed by either general or differential space absorption.
>
> In summary, we conclude that though readily observed obscuring matter effectively conceals some of the galactic star clouds in the central region, and even hides the center itself, the Galaxy is completely transparent at the borders of these obscuring nebulae.[48]

Unfortunately, Shapley's data on extragalactic nebulae were incomplete.

Trumpler's results,[49] which we mentioned briefly before, were published in 1930. Apparently his research had attracted the attention of several astronomers and the publication was eagerly awaited, as is indicated in a letter from Trumpler to R. Aitken.[50] There is little wonder 'at the excitement and interest in Trumpler's research since he finally provided definite proof of the existence of diffuse obscuring matter.

Trumpler for several years had studied galactic clusters, i.e., open clusters normally found close to the galactic equator. While doing his research, Trumpler developed two methods of determining the distance to the clusters. The first method involved the apparent magnitudes of stars, whose absolute magnitudes he inferred from plots of the H-R diagram. The second

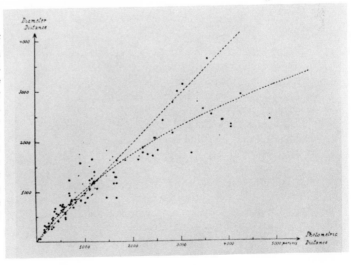

A comparison of the distances of open clusters determined from apparent magnitudes and spectral type (abscissa) with those determined from angular diameters (ordinate). The dotted straight line indicates the expected relation if no absorption is present. The dotted curve gives the relation for a general absorption of 0^m7/kpc. (Courtesy of the *Publications* of the Astronomical Society of the Pacific, 42, 214, 1930)

James Lick, donor of Lick Observatory. (Lick Observatory photograph)

The site picked for Lick Observatory. (Lick Observatory photograph)

method depended upon the inverse relation between distance and apparent size of a cluster.

A discrepancy between the distances determined by the two methods became apparent. The farthest clusters have angular diameters larger than expected, when the nearer clusters are used as standards. Trumpler searched for possible observational errors that would explain the apparent increase in diameter with distance from the solar system, but none of the errors could account for the behavior of the clusters. Only two possible alternatives were then left, "either to admit an actual change in the dimensions of open clusters with increasing distance or to assume the existence of an absorption of light within our stellar system."[51] He definitely favored the second alternative, and added that observations of the colors of the clusters supported his view. An absorption coefficient of 0^m67/kpc for photographic magnitudes eliminated the discrepancy between distances determined by the two methods outlined above.

Trumpler's research was more extensive and complete than previous studies carried out to determine the amount of general absorption. This does not mean, however, that he did not benefit from the efforts of the earlier investigators. Although studies prior to 1930 had not been conclusive, many of them indicated that absorption could not be ruled out completely; hence an atmosphere was created where the concept

of general absorption was easily acceptable. Conjectures, such as those by Curtis about the region of avoidance, also apparently persuaded some astronomers to accept the existence of general absorption, as is indicated by Oort's statements.

Trumpler faced several problems relating to the observations of globular clusters. The main obstacle to acceptance of the value for absorption was the lack of reddening in globular clusters. Trumpler had no trouble providing an explanation.

It is natural to interpose here the question why such an absorption of light should not have been discovered in the discussion of the diameters of globular clusters which are much more distant, and how it is possible that we still find small color-indices [blue colors] in some globular clusters . . . despite of their great distances. There is only one way which seems to lead out of the dilemma: the hypothesis that the absorbing medium, like the open clusters, is very much concentrated toward the

"FRASERIUS ET HUBBARDUS ANTIQUUS"

galactic plane. . . . Perhaps this absorbing material is related to interstellar calcium or to diffuse nebulae which are also strongly concentrated to the galactic plane.[52]

The rest of his paper described the distribution in space of open clusters and discussed possible structural models of our Galaxy.

Trumpler's results were very important to galactic astronomy. The theories of Kapteyn and Shapley had to be reconsidered. Absorption explained why Kapteyn had detected a decrease in stellar number densities with increasing distance from the Sun, and why the Sun seemed to be the center of the system. Kapteyn's doubts in 1909 about the reality of the observed distribution turned out to be justified—the distribution was not real, but only an apparent one produced by the absorption. The size of our Galaxy as determined

Frazier (*left*) was a site surveyor for Lick. Hubbard (*mounted*) was an old-timer who lived in the area chosen for the observatory. He had a wooden leg, and he sported six notches on his gun. (Lick Observatory photograph)

Robert J. Trumpler: 1886–1956

Robert Trumpler was born in Switzerland in the year 1886, third in a family of ten children. He received his first education in Zurich, Switzerland, where his scholastic record was not outstanding. Later, when he entered the Gymnasium, his interest was aroused, especially in astronomy, and his studies improved greatly.

Trumpler's father, a successful businessman, did not approve of astronomy as a career, for reasons of economics, so that young Trumpler could only make a hobby of astronomy, and he turned to business training. The arrangement did not succeed, however, and he entered the University of Zurich to study astronomy, physics, and mathematics. He completed his education at Göttingen, Germany, where he received his doctorate in 1910. The following year he accepted the position of Astronomer of the Swiss Geodetic Commission at Basel.

In 1913, at a meeting of the Astronomische Gesellschaft, Trumpler met Frank Schlesinger, of the Allegheny Observatory; they discussed common interests in research. A few months later, Schlesinger offered Trumpler a place at the observatory. Trumpler was delighted to accept, since the Geodetic Commission offered limited opportunities. Unfortunately, before the final details were worked out and Trumpler could arrange to travel to America, war broke out in Europe. Although Switzerland was not immediately involved in the conflict, the entire country was mobilized in preparation for a possible invasion. As every man in Switzerland is affected by general mobilization, Trumpler soon found himself stationed in the Alps as officer in the Swiss Militia.

He almost lost his opportunity to go to the United States, since Schlesinger could not leave his position unfilled in-

A stagecoach on the way up Mount Hamilton to Lick Observatory. (Lick Observatory photography)

definitely. Luckily, a neutral country during a war can be useful to both sides, and Switzerland's neutrality was respected. Since there was no longer a need for total mobilization, Trumpler's appeal for leave to go to the United States was granted.

Under Schlesinger, Trumpler began to investigate open clusters. His research in that field brought him to the attention of the director of the Lick Observatory. In 1918 he accepted an invitation to come to Lick, where he stayed for the rest of his life. He continued his detailed studies of open clusters, which eventually led him to the definite confirmation of an absorbing medium in interstellar space. That achievement was a major contribution to galactic astronomy.

Robert Trumpler's comparison of the distribution of galactic clusters and globular clusters. The region of the galactic clusters (Kapteyn universe) is indicated by the shaded oval. The positions of the globular clusters are indicated by dots. The two shaded circles represent the Magellanic Clouds. (Lick Observatory Bulletin, 1930)

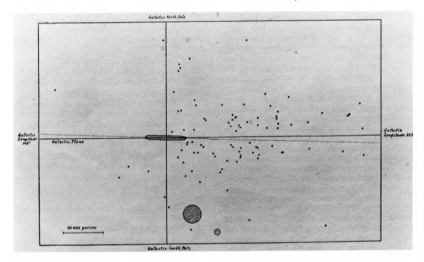

by Kapteyn could no longer be accepted.

Shapley's model was not made invalid, but it was modified by Trumpler's results. The apparent distances of the farthest globular clusters at low galactic latitudes were affected tremendously by absorption. Eventually the diameter of Shapley's model was reduced by a factor of three, and the system of globular clusters was found to be nearly spherical instead of ellipsoidal. The region of avoidance of the globular clusters and the spiral nebulae was easily explained as a result of the flat distribution of absorbing matter.

Trumpler's investigation definitely had a great influence on galactic theories but, amazingly, he seems to have been unaware of all its consequences. He still was faithful to Kapteyn's model even though the absorption results spelled its doom. Trumpler even sharply attacked Shapley's model.

Lick Observatory and the astronomers' residence *c.* 1900. (Lick Observatory photograph)

Perhaps the reason Trumpler was faithful to Kapteyn's model was his belief that our Galaxy is a spiral nebula and that Kapteyn's diameter corresponded to those determined by Hubble, in 1924, for the Andromeda Nebula and Messier 33. Hubble determined diameters of these nebulae as 13 kpc and 4.6 kpc, respectively; Kapteyn estimated our Galaxy as being between 10 and 15 kpc in diameter, whereas Trumpler had found open clusters over a region between 10 and 12 kpc in diameter. Trumpler could not reconcile Shapley's great distance with his belief that our Galaxy is a spiral nebula. As the next section will show, Shapley's corrected distances were compatible with the diameters of spiral nebulae, once their correct distances were determined.

94

Trumpler did not accept Oort's recent results as evidence for a longer galaxy.

The evidence for galactic rotation around a center in the Sagittarius region is mainly based on a second harmonic in the observed radial velocities of stars and interstellar calcium. What is observed, of course, is a differential effect and its interpretation necessarily requires certain assumptions about the nature of rotational motion; furthermore, the observations furnish only the direction of the center of rotation, but not its distance.[53]

He also refused to accept Shapley's conclusion that globular clusters are intimately connected with the Milky Way system, and noted a valid objection:

While the globular cluster system appears nearly spherical in shape there can be no question that the Milky Way system is much flattened. . . . The majority of globular clusters thus lie outside of the star stratum of our Milky Way system and should in this sense be extra-galactic systems. . . .[54]

Trumpler did not repudiate all connection between the globular clusters and the Milky Way; instead he

The road up Mount Hamilton to Lick Observatory in 1907. (Lick Observatory photograph)

relegated them to the outer regions of what he called a supercluster containing one spiral nebula, the two Magellanic Clouds, and hundreds of globular clusters.

His model did not gain even momentary acceptance, because the consequences of absorption were soon realized.

The interment of James Lick beneath the pier of the 36-inch telescope. (Lick Observatory photograph)

Epilogue

for Section Two

Section Two has resolved the conflict between the Kapteyn Universe and Shapley's scheme of our Galaxy. In doing so, we have explored several topics of astronomical research that were born in the 1920s and others that were spurred to greater activity by the developments of that period. Galactic rotation, general absorption interstellar absorption lines and galactic clusters are the main areas of interest discussed in Section Two.

Photography applied to astronomical research produced many of the great strides in our knowledge of the heavens; the birth of modern astronomy would have been impossible without the photographic plate. Spectroscopy, radial velocities, proper motions, and studies of stellar clusters all depended heavily upon photography. Without the basic data provided by photography, modern theories of our Galaxy would not have developed so rapidly.

The Kapteyn Universe is an example of a theory developed through mathematically sophisticated use of the photographically-determined basic data. Unfortunately, Kapteyn was misled by inaccurate determinations of the amount of interstellar absorption.

Lindblad and Oort developed a new theory for the dynamics of our Galaxy. They postulated that stars are revolving about a common center. The stars in our neighborhood move in nearly circular orbits, while other stars move in highly elliptical orbits. The model developed by the two men, independently, explained the nature of the high-velocity stars, which had previously presented a difficult puzzle to astronomers. The model also predicted that the effects of differential galactic rotation would be visible in the behavior of radial velocities and proper motions of stars in our vicinity. Oort searched for the effects and found them. From the data he gathered, Oort was able to predict the distance to the center of our Galaxy, which was much less than the distance determined by Shapley from the study of globular clusters. The resolution of this discrepancy lay in the determination of the amount of interstellar absorption.

Trumpler, in his study of galactic clusters, provided the first definite proof of the existence of a generally absorbing interstellar medium. He found systematic divergences that could be explained by assuming that a thin layer of absorbing material existed in the galactic plane. Although Trumpler apparently did not recognize the fact, the absorption he found explained the discrepancies between Kapteyn's and Shapley's models of our Galaxy, as well as the discrepancies between Oort's and Shapley's distances to the galactic center. With the determination of the amount of interstellar absorption, all the pieces of the puzzle fell into place; contradictions were reconciled.

The early studies of our Galaxy in the 1920s and 1930s were important and provided the first basic models capable of explaining stellar observations, but they were only the beginning. Today we have a better understanding of our Galaxy, although not a complete understanding by any means. Radio observations in the 1950s confirmed what people had been speculating about for years—our Galaxy shows indications of having spiral structure. Studies of the polarizing and absorbing abilities of the interstellar medium as a function of wavelength of light have increased our understanding of the types of grains that must exist in space. Several models for the grains have been proposed, but no one model seems to explain all the observations. The full nature of the interstellar grains that cause absorption is still not definite. Also, Trumpler's model of a uniform interstellar absorbing medium is now known to be only a rough approximation to what actually exists. There is evidence that the absorbing material occurs in small clouds, with relatively clear space in between. However, over large distances, the absorption effects of the many clouds are averaged together so that the end result does not differ greatly from that of a completely uniform distribution. Certain areas of the sky, however, are subject to greater amounts of absorption than others. The fine details of the early galactic models, proposed in the 1920s and 1930s, resulted from more sophisticated techniques developed in later years.

References

Chapter 1

1 R. L. Waterfield, A Hundred Years of Astronomy (New York: Macmillan, 1938): 318.
2 C. Easton, "A New Theory of the Milky Way," Astrophys. J. 12 (1900): 36.
3 Ibid., 157.
4 C. Easton, "A Photographic Chart of the Milky Way and the Spiral Structure of the Galactic System," Astrophys. J. 37 (1913): 105–118.
5 Ibid., 118.
6 Ibid., 116.
7 A. Pannekoek, A History of Astronomy (London: George Allen and Unwin Ltd., 1961): 470.
8 J. C. Kapteyn, Plan of Selected Areas (Groningen: Hoitsema Bros., 1906).
9 J. C. Kapteyn, "First Attempt at a Theory of the Arrangement and Motion of the Sidereal System," Astrophys. J. 55 (1922): 302–328.
10 Ibid., 302
11 J. C. Kapteyn, "On the Absorption of Light in Space," Astrophys. J. 29 (1909): 47.
12 H. Shapley, "Studies Based on the Colors and Magnitudes in Stellar Clusters," Astrophys. J. 45 (1917): 118–141 and 164–181; 46 (1917): 64–75; 48 (1918): 89–124, 154–181, and 270–294; 49 (1919): 24–41, 96–107, 249–265, and 311–336; 50 (1919): 42–49 and 107–140.

Chapter 2

1 G. Stromberg, "The Motions of the Stars and the Existence of a Velocity Restriction in a Universal World-Frame," Sci. Mon. 9 (1924): 470.
2 W. S. Adams and A. Kohlschutter, "The Radial Velocities of One Hundred Stars with Measured Parallaxes," Astrophys. J. 39 (1914): 348.
3 W. S Adams and A. Kohlschutter, "The Motions in Space of Some Stars of High Radial Velocities," Astrophys. J. 49 (1919): 183.
4 R. L. Waterfield, A Hundred Years of Astronomy (New York: Macmillan, 1938): 318, 324–325.
5 J. H. Oort, "Some Peculiarities in the Motions of Stars of High Velocity," Bull. Astron. Inst. Neth. 1 (1922): 133–137.
6 J. H. Oort, "The Development of Our Insight in the Structure of the Galaxy Between 1920 and 1940," talk at N. Y. Academy of Sciences, September 1971.
7 J. H. Oort, The Stars of High Velocity (Thesis, Rijks-Universiteit te Groningen, 1926).
8 Ibid., 63.
9 Ibid., 63–64.
10 Ibid., 64.
11 Ibid., 64–65.
12 Ibid., 65.
13 Ibid., 66.
14 Ibid.
15 Private communication, J. H. Oort to B. Lindblad, 8 April 1927 (in possession of P. O. Lindblad).
16 J. H. Oort, "Observational Evidence Confirming Lindblad's Hypothesis of a Rotation of the Galactic System," Bull. Astron. Inst. Neth. 3 (1927): 275–282.
17 Ibid., 279.
18 J. H. Oort, "Investigations Concerning the Rotational Motion of the Galactic System, Together with New Determinations of Secular Parallaxes, Precession and Motion of the Equinox," Bull. Astron. Inst. Neth. 4 (1927): 88.
19 Private communication, J. S. Plaskett to J. H. Oort, 28 November 1927 (Leiden Observatory).
20 J. S. Plaskett, "The Rotation of the Galaxy," Mon. Not. R. Astron. Soc. 88 (1928): 395–403.
21 Private communication, J. H. Oort to J. S. Plaskett, 22 December 1927 (Leiden Observatory).

Chapter 3

1 E. E. Barnard, The Bruce Photographic Telescope," Astrophys. J. 21 (1905): 46.
2 Ibid., 47.
3 E. E. Barnard, On a Nebulous Groundwork in the Constellation Taurus," Astrophys. J. 25 (1907): 221.
4 E. E. Barnard, "On a Great Nebulous Region and on the Question of Absorbing Matter in Space and the Transparency of the Nebulae," Astrophys. J. 31 (1910): 13.
5 E. E. Barnard, "On the Dark Markings of the Sky with a Catalogue of 182 Such Objects," Astrophys. J. 49 (1919): 1.
6 J. C. Kapteyn, "On the Absorption of Light in Space," Astrophys. J. 29 (1909): 3.
7 J. C. Kapteyn, "On the Absorption of Light in Space," Astrophys. J. 30 (1909): 163–196.
8 Kapteyn originally reported a value for absorption of about 0^m5/kpc. He discovered an error in his average distances of a factor of 1.7, which he reported to G. E. Hale (see private communication, J. C. Kapteyn to G. E. Hale, 7 November 1909, in the Hale microfilm collection, Pasadena). The absorption he originally reported had to be divided by this factor to be valid.
9 Private communication, J. C. Kapteyn to G. E. Hale, 14 January 1913 (Hale Collection).
10 Private communication, G. E. Hale to J. C. Kapteyn, 6 January 1914 (Hale Collection).
11 Among the investigations that Kapteyn reviewed were: E. A. Fath, "The Integrated Spectrum of the Milky Way," Astrophys. J. 36 (1912): 362–367; E. S. King, "Photographic

Magnitudes of 153 Stars," *Ann. Havard Coll. Obs. 8* (1912): 157–186; H. H. Turner, "Interstellar Space," *Mon. Not. R. Astron. Soc. 99* (190): 61–71; E. C. Pickering, *Harvard Circular No. 170*; F. H. Seares, *Mt. Wilson. Cont. 81* (1913); W. S. Adams, "Note on the Relative Intensity at Different Wavelengths of the Spectra of Some Stars Having Large and Small Proper Motion," *Astrophys. J. 39* (1914): 89–92; and several unpublished works by Barnard, Hertzsprung, and van Rhijn.

12 J. C. Kapteyn, "On the Change of Spectra and Color Index with Distance and Absolute Brightness, Present State of the Question," *Astrophys. J. 40* (1914): 187–204.

13 Private communication, J. C. Kapteyn to G. E. Hale, 23 September 1915 (Hale Collection).

14 H. Shapley, "Studies Based on the Colors and Magnitudes in Stellar Clusters, First Part: The General Problem," *Astrophys. J. 45* (1917): 130.

15 Private communication, H. Shapley to F. Moulton, 7 January 1916 (Harvard University Archives).

16 H. Shapley, "Globular Clusters and the Structure of the Galactic System," *Publ. Astron. Soc. Pac. 30* (1918): 50.

17 Private communication, G. E. Hale to H. Shapley, 14 March 1918 (Harvard University Archives).

18 Private communication, H. N. Russell to H. Shapley, 13 March 1919 (Harvard University Archives).

19 H. D. Curtis, "Absorption Effects in the Spiral Nebulae," *Proc. Nat. Acad. Sci. USA 3* (1917): 678.

20 Private communication, H. N. Russell to Shapley, 13 March 1919.

21 *Ibid.*

22 H. Shapley and M. B. Shapley, "Studies Based on the Colors and Magnitudes in Stellar Clusters, Fourteenth Paper: Further Remarks on the Structure of the Galactic System," *Astrophys. J. 50* (1918): 118.

23 Private communication, H. N. Russell to H. Shapley, 20 March 1919 (Princeton University Archives).

24 Private communication, H. Shapley to H. N. Russell, 18 May 1919 (Princeton University Archives).

25 Private communication, H. N. Russell to H. Shapley, 9 June 1919 (Harvard University Archives).

26 J. C. Kapteyn, "Absorption of Light," *Astrophys. J. 29* (1909): 48, 54.

27 V. M. Slipher, "Peculiar Star Spectra Suggestive of Selective Absorption of Light in Space," *Lowell Obs. Bull. 2,* 1 (1909): 2.

28 Private communication, J. C. Kapteyn to V. M. Slipher, 30 October 1909 (Kapteyn Laboratorium, Groningen).

29 J. S. Plaskett and J. A. Pearce, "The Problems of Diffuse Matter in the Galaxy," *Publ. Dominion Astrophys. Obs. 5* (1933): 169.

30 *Ibid.*

31 O. Struve, "On the Calcium Clouds," *Popular Astronomy 33* (1925): 639–653; *34* (1926): 1–14.

32 A. S. Eddington, "Diffuse Matter in Interstellar Space," *Proc. R. Soc. Lond. Ser. A, 3* (1926): 424.

33 *Ibid.,* 445.

34 O. Struve, "Interstellar Calcium," *Astrophys. J. 65* (1927): 174–175.

35 *Ibid.,* 197.

36 *Ibid.,* 198.

37 *Ibid.*

38 O. Struve, "Further Work on Interstellar Calcium," *Astrophys. J. 67* (1928): 390.

39 *Ibid.,* 383.

40 Private communication, J. H. Oort to O. Struve, 21 November 1928 (Leiden Observatory).

41 Private communication, O. Struve to J. H. Oort, 15 December 1928 (Leiden Observatory).

42 B. P. Gerasimovic and O. Struve, "Physical Properties of a Gaseous Substratum in the Galaxy," *Astrophys. J. 69* (1929): 31.

43 Private communication, O. Struve to J. H. Oort, 29 August 1929 (Leiden Observatory).

44 J. S. Plaskett and J. A. Pearce, "The Motions and Distribution of Interstellar Matter," *Mon. Not. R. Astron. Soc. 90* (1930): 267.

45 Private communication, 23 December 1969.

46 J. H. Oort, "Observational Evidence Confirming Lindblad's Hypothesis of a Rotation of the Galactic System," *Bull. Astron. Inst. Neth. 3* (1927): 281.

47 Private communication with Nicholas Mayall, 13 April 1970.

48 H. Shapley, "Studies of the Galactic Center IV. On the Transparency of the Galactic Star Clouds", *Proc. Nat. Acad. Sci. USA 15* (1929): 175, 177.

49 R. Trumpler, "Preliminary Results on the Distances, Dimensions and Space Distribution of Open Star Clusters," *Lick Obs. Bull. 14* (1930): 154–188.

50 Private communication, R. Trumpler to R. Aitken, 28 January 1930 (Lick Observatory).

51 Trumpler, "Preliminary Results," 163.

52 *Ibid.,* 166–167.

53 *Ibid.,* 186–187.

54 *Ibid.,* 187.

THREE

Extragalactic Nebulae

Contents

of Section Three

Prologue

for Section Three

Edwin Hubble with his cat, Nikolus Copernicus. (*Colliers* Magazine, 1949)

The early arguments about the size and structure of the universe were investigated and brought into dramatic focus in the Curtis-Shapley debate of 1920. Two issues arose from the debate: the nature of the Milky Way (the subject of Section Two) and the nature of certain nebulous objects, in particular, the spiral nebulae (the subject of this section).

The nature of the spirals could be settled only by determining distances to them. If those distances proved to be extremely large, then the spirals must be not only external to the Milky Way but also possibly comparable to it.

Until 1925, the only apparently reliable distance estimates, based on studies of proper motions, indicated that spirals could not be at the great distances required by the island–universe theory. After 1925, however, distances to the spirals were found from studies of the stars within them. These estimates clearly showed the spirals to be at enormous distances from the Sun. Hence, a conflict arose between these two sets of observations: the proper-motion results of Adriaan van Maanen and the Cepheid results of Edwin Hubble.

Although most astronomers were quickly convinced that Huble's results were correct, the controversy continued until 1935, when Hubble demonstrated that van Maanen's work was subject to systematic errors and therefore invalid. The island-universe theory became accepted—180 years after Kant had initially postulated it. Interestingly, the interval from Kant to Hubble was longer than that from Copernicus to Newton! (The interval between Copernicus' *De Revolutionibus* and Newton's *Principia* was 144 years, whereas that between Kant's *Universal Natural History* and Hubble's discovery of Cepheids in spirals was 170 years.)

The Spiral galaxy NGC 4565 seen edge-on. (Hale Observatories photograph)

CHAPTER 1

Radial Motions of Spiral Nebulae

Discovery of Radial Velocities of Spirals

What has become one of the most important discoveries in modern astronomy—that spiral nebulae have large radial velocities and that almost all are moving away from the Milky Way—was initially made by Vesto M. Slipher in 1912.

The motivation that led to this discovery, however, came from Percival Lowell. Lowell was director of the observatory at Flagstaff, Arizona, when Slipher joined the staff in 1901.[1] Lowell had founded the observatory in 1894, using a borrowed 18-inch telescope. Observations of Mars began immediately, since that planet was one of Lowell's primary interests. He wrote many articles about Mars and generated a great deal of public interest. Because of Lowell's driving personality, the observatory grew rapidly. In 1897, a 24-inch Alvan Clark refractor was installed, and in 1900 a Brashear spectrograph was ordered.

104

Since Lowell was also interested in the Chamberlain–Moulton hypothesis that spiral nebulae evolve into solar systems, Slipher was given the task of installing the spectrograph and investigating the spirals for evidence of rotation. The measurements were extremely difficult to make because the nebulae are so faint. For this reason Slipher began by measuring the rotations of planets—a program in which he was remarkably successful. By 1912, he had finally obtained spectra of the Andromeda Nebula that showed Doppler shifts in the spectral lines.[2] The results were startling. They showed the Andromeda Nebula to be approaching the Sun with a velocity of 300 km/sec—the greatest velocity that had been observed in astronomy.

Slipher continued his observations and by 1914, when he first published his results, he had obtained Doppler shifts for 14 spirals. His announcement, made at the Evanston meeting of the American Astronomical Society,[3] was greeted with a standing ovation—a remarkable event for a scientific meeting. This work also won for him the gold medals of the Paris Academy, the Royal Astronomical Society, and the Astronomical Society of the Pacific. Soon after this announcement, several other astronomers obtained results that verified Slipher's. Nevertheless, the work was continued almost solely by Slipher. By 1925, of the 45 known velocities, all but 5 were due to Slipher (although 10 were independently verified).

Other astronomers were also trying to take good spectra of spiral nebulae; although Slipher was not the first to succeed, he was the first to perfect the techniques that made accurate measurements of the spectral lines possible. Like Lowell, he did not have the benefit of graduate study (both began as largely self-taught amateurs), but he possessed a strong background in mathematics and a good understanding of technical problems. Slipher did receive a Ph.D., although under somewhat unusual circumstances. After publishing a paper on the spectrum of Mars and passing an oral exam on it, he was awarded a doctorate by the University of Chicago, which waived the usual requirements for gradate work.

Implications for the Nature of the Nebulae: I

Since the spirals had such extraordinarily large radial velocities, many astronomers were convinced that they probably could not be contained within the Milky Way. On 14 March 1914, just weeks after Slipher's original announcement, E. Hertzsprung wrote to him:

A portrait of Percival Lowell, founder and director (1894–1916) of Lowell Observatory at Flagstaff, Arizona. (Yerkes Observatory photograph)

My harty [sic] congratulations to your beautiful discovery of the great radial velocity of some spiral nebulae.

It seems to me, that with this discovery the great question, if the spirals belong to the system of the milky way or not, is answered with great certainty to the end, that they do not.[4]

Slipher himself suggested that the spirals were outside the Milky Way, when he postulated the theory that the Milky Way galaxy was drifting with respect to the spirals.[5] He formulated the idea to explain why some of the spirals were apparently approaching the Milky Way and others (the vast majority) were receding from it. Even the public became interested in the problem, and the *New York Times* published an article in 1921,[6] suggesting that the large radial velocities implied a large distance to the spiral nebulae.

Even though some astronomers appear to have been convinced by Slipher's results, these findings did not prove the island-universe theory, since they could not yield the distances to spirals. Without knowledge of these distances, the spirals' relation to the Milky Way could not be determined. All that could be inferred from the radial-velocity results at that time was that spirals seemed to exhibit peculiar properties.

Although Slipher's radial velocities did not yield the nature of the spirals, they later profoundly influenced astronomy, as we shall discuss in detail in Section Four.

The spiral nebula in Andromeda— Messier 31 (NGC 224). (Hale Observatories photograph)

Vesto Melvin Slipher; 1875-1969

Daniel Clarke and Hannah (App) Slipher had two sons, both of whom became famous astronomers, made important studies of the planets, and directed the Lowell Observatory—Vesto M. and Earl C. (1883–1964) Slipher. It was the older son, however, who perfected techniques in spectroscopy and achieved great advances in galactic astronomy.

V. M. Slipher earned his A.B. (1901), M.A. (1903), and Ph.D. (1909) degrees at the University of Indiana, and received honorary degrees from the University of Arizona (Sc.D., 1923), the University of Indiana (LL.D., 1929), the University of Toronto (Sc.D., 1935), and Northern Arizona University (Sc.D., 1965).

Slipher's main research contributions were in spectroscopy, where he pioneered instrumental techniques as well as made major discoveries. His research can be divided into three main areas: planetary atmospheres and rotations, diffuse nebulae and the interstellar medium, and rotations and radial velocities of spiral nebulae.

He had determined the rotation periods of Venus and Mars by 1903, and of Jupiter, Saturn, and Uranus by 1912. He was also instrumental in discovering part of the chemical composition of the Jovian atmosphere. The Royal Astronomical Society awarded Slipher the Gold Medal in 1933 for his work on planetary spectroscopy.

In 1912 he showed that some diffuse nebulae shone by reflected light, thus clearly demonstrating the existence of particulate matter in interstellar space.

The most significant area of Slipher's research, however, dealt with spiral nebulae. He was the discoverer of radial and rotational velocities of spirals and a leader in such research.

Other areas of Slipher's research included the determination of radial velocities of globular clusters, spectroscopic studies of comets and aurorae, and observations of bright lines and bands in night-sky spectra.

Slipher was also an unusually competent administrator; indeed, in recognition of his ability as an administrator as well as a researcher, in 1935 the Astronomical Society of the Pacific awarded him the Bruce Medal.

He received his first experience in administration in 1915 when Percival Lowell made him assistant director of the observatory. He became acting director upon Lowell's death in 1916 and continued in that capacity until he was made director in 1926, a post he held until 1952, when he became director emeritus.

During his directorship he supervised the search for a trans-Neptunian planet, which culminated in 1930 in the discovery of Pluto, by Clyde Tombaugh, a staff member at Lowell.

Slipher's other administrative experience included serving as president of the Commission on Nebulae (No. 28) of the International Astronomical Union (1925 and 1928), and vice-president of the American Association for the Advancement of Science (1933).

CHAPTER 2

Rotational Motions of Spiral Nebulae

Spectrographic Results

Another discovery by Slipher that had a more pronounced effect on the question of the nature of spirals was that they are rotating.[1] This result, also confirmed by Max Wolf,[2] indicated that it might be possible to detect proper motions within the spirals, and hence the distances to them.

Proper-Motion Results

The question of proper motions in spiral nebulae was first critically discussed by Curtis in 1914.[3] Since the Lick Observatory, using the famous Crossley Reflector, began in that year to repeat the program of nebular photography begun 16 years before by James Keeler, astronomers hoped to determine proper motions or internal changes in the spirals. According to

The spiral nebula Messier 101. Left, an 1851 drawing by S. Hunter using Lord Rosse's 72-inch telescope. Right, a photograph by H.D. Curtis, c. 1915. (*Adolfo Stahl Lectures in Astronomy*, Stanford University Press: San Francisco, 1919)

Curtis, this was important because "a knowledge of the proper motions or of any rotational movements which these bodies may have would be of great value in investigations as to the size and distance of the nebulae, and therefore as to their place in the structure of the visible universe."[4]

Although the program was not complete at the time of Curtis' paper, he thought that from the observations already made he could draw the general conclusion that "in this average interval of thirteen years it has not been possible to detect any evidence of internal movement, rotary or otherwise, in the nebulae measured."[5]

Curtis also concluded, since Wolf and Slipher had reported rotations in spirals from their spectrographic studies, that the spirals were at a great distance. He

Above and *below:* Two views of one of the stereocomparators at Mount Wilson used by van Maanen to measure internal proper motions in spiral nebulae. (by R. Berendzen, 1972)

wrote, "As the spirals are undoubtedly in revolution—any other explanation of the spiral form seems impossible—the failure to find any evidences of rotation would indicate that they must be of enormous actual size, and at enormous distances from us."[6]

Other, far more influential, evidence soon became available, however, due to the efforts of Adriaan van Maanen. In 1912, van Maanen was appointed to the staff at Mount Wilson where he began measuring proper motions and parallaxes of stars. He was ideally suited for this work not only because his dissertation (*The Proper Motions of 1418 Stars in and near the Clusters h and χ Persei*, Utrecht, 1911) had been on proper motions, but also because his work at Yerkes

110

during 1911-1912 had involved making such measurements.

By 1915, motions had been discovered in several nebulae, including rotational and radial motions in some spirals. It is thus not surprising that, in December of that year, George W. Ritchey asked van Maanen to determine if proper motions existed in the spiral M101. Ritchey suggested that two of his plates, which had been taken five years apart (1910 and 1915) with Mount Wilson's 60-inch reflector, be studied with the stereocomparator. Van Maanen found some evidence of internal motion, but he felt that more plates should be examined. He asked Curtis for the additional plates, since Curtis had conducted a program at Lick that had specifically looked for proper motions of nebulae. Van Maanen received the plates in early 1916 and used them to confirm his earlier results; the subsequent papers were sent to the *Astrophysical Journal* in March and to the National Academy of Sciences in June.[7]

An early portrait of Adriaan van Maanen. (*Porträtgallerie der Astronomischen Gesellschaft*, Budapest, 1931)

Van Maanen used a blink stereocomparator to examine the plates in pairs—a method he had previously used to determine proper motions for stars. The method consisted of alternately looking at two plates of the same region (the plates being taken several years apart) to see if any of the images had moved. His results showed a mean annual rotational motion of $0\rlap{.}{''}022$ (left-handed) and a mean annual radial motion of $0\rlap{.}{''}007$ (outward). These observations yielded three important conclusions: 1. spiral nebulae are undoubtedly rotating; 2. the motion of objects in spirals is outward, i.e., as if they were unwinding; 3. the large rotations require the spirals to be close to or even inside the Milky Way.

Close-up of the sign on the front of the stereocomparator. (R. Berendzen, 1972)

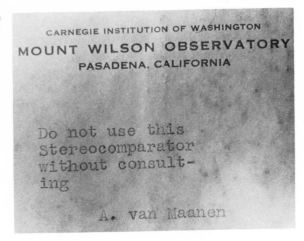

CARNEGIE INSTITUTION OF WASHINGTON
MOUNT WILSON OBSERVATORY
PASADENA, CALIFORNIA

Do not use this
stereocomparator
without consult-
ing

A. van Maanen

Implications for the Nature of the Nebulae: II

This last conclusion is obvious from a consideration of the simple relations among proper motion (μ), distance (d), and rotational velocity (v):

$$v \sim \mu d$$

Since μ is not very small, a reasonably small v (say, 200 km/sec) requires a small distance. Also, if d is taken to be very large (commensurate with the island-universe theory), the value for v becomes excessive.

In 1921, van Maanen published his measurements for three more spirals (M33, M51, M81).[8] By mid-1916 he had made some preliminary determinations of motions in M81 which John C. Duncan had reported to Slipher, noting that they were in agreement with those obtained for M101.[9] Van Maanen wrote to Shapley in June, 1921, informing him that the work on M81 was completed and that it showed "motion like the others."[10]

The 60-inch reflector at the Mount Wilson Observatory. (Hale Observatories photograph)

Van Maanen often described his work to Hale through letters. In one written in 1917[11]—only a year

TABLE 2.1 Summary of van Maanen's results for
internal motions in seven spirals

Object	Date Published[16]	Mean $\mu_{rot}["/yr]$	Mean $\mu_{rad}["/yr]$
M101	1916	0.021±0.001	0.003±0.001
M33	1921	0.020±0.003	0.006±0.002
M51	1921	0.019±0.001	0.008±0.001
M81	1921	0.020±0.004	0.017±0.003
NGC 2403	1922	0.015±0.001	0.014±0.001
M94	1922	0.020±0.002	0.010±0.002
M63	1923	0.019±0.001	0.004±0.001
M33	1923	0.020±0.001	0.003±0.001

after his paper on M101—he transmitted his preliminary result for the parallax of the spiral M51. His early finding was $p = 0.''000 \pm 0.''010$—obviously, he was attempting a measurement at the limit of the precision of his data.

Similarly, his measurements of the great spiral in Andromeda (M31) were not significant.[12] In late 1917, he wrote to Hale about his findings:

I have just finished the parallax of the Andromeda Nebula . . . $[p] = +0.''004 \pm 0.''005$. So that we do not know as yet if this is an island universe! I think it will be good to try to take some plates of the regions where there are a large number of condensations, in the hope of eventually finding some proper motion.[13]

Thus, by 1917, van Maanen began to realize or, at least, to admit the implications of his measurements for the island-universe theory; he discussed parallaxes with colleagues and even specifically mentioned the theory, in the letter cited above. But he did not publicly state the implication until 1921, in the paper on M33 (the paper was communicated in November, 1920), when he asserted that the internal motions "would raise a strong objection to the 'island universe' hypothesis."[14] He argued that "if comparable to our own system, spiral nebulae would be so distant that no ordinary velocities of their constituent parts would yield measurable displacements in the short interval of a few years."[15] He concluded this paper by arguing that if M33 were comparable to our galaxy, it must be remote, and the measured displacements would then indicate a rotation on the order of the velocity of light.

Van Maanen continued making measurements until 1923, when he had exhausted his supply of early–

A 1910 photograph of Messier 101 made by G.W. Ritchey at Mount Wilson. (Hale Observatories photograph)

epoch plates; by then he had results for seven spirals. In 1923, in the last paper he published giving new data on motions in spirals, he concluded that:[17]

1. Errors due to the telescope, the quality of the plates, the measuring instrument, or the measurer could not have caused the findings; the measurements represented real internal motions.
2. The results agreed with Jeans' theoretical work (see Chapter 3).
3. The parallaxes of the spirals are of the order of a few thousandths of a second of arc, corresponding to distances of only a few thousand parsecs.
4. The sizes of the spirals are small compared with that of the Milky Way.

Influence on the Great Debate

In 1919 Shapley published a paper in which he summarized the arguments for and against the island-universe theory.[18] He listed van Maanen's measure-

114

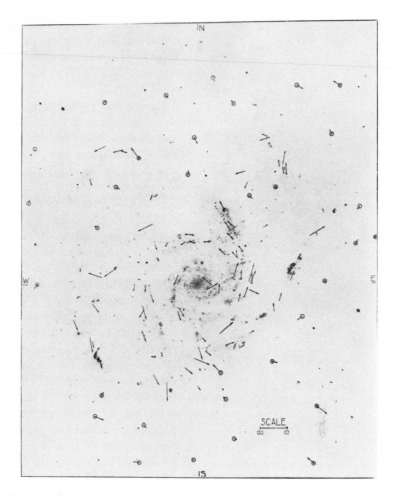

ments of internal motions in M101 as one of several pieces of evidence unfavorable to the theory, and he concluded that the evidence supporting the theory was unconvincing.

The published version of the Curtis-Shapley debate contains Shapley's arguments against the theory; he listed van Maanen's results for M101 and M81 as the last of his three negative arguments. In the debate itself, however, Shapley made virtually no mention of the theory. Shapley later maintained that he failed to discuss the issue in the debate simply because he was not especially interested in it and did not think that it was pertinent to the topic.[19] Nevertheless, in a letter to Russell on 31 March 1920, he wrote that in the forthcoming debate he would not say much about spirals because "I have neither time nor data *nor very good argument*"[20] (emphasis added). This statement probably represents some lingering doubt of Shapley's, since he was apparently somewhat skeptical of van Maanen's results from the beginning, even though in

A plate from van Maanen's 1916 paper in which he described the discovery of internal motions in the spiral Messier 101. The arrows indicate the magnitudes and directions of the mean annual motions. The comparison stars are enclosed in circles. (*Astrophysical Journal, 44*, 1916, University of Chicago Press)

his later years he said that he had believed them whole-heartedly. For instance, in 1917 he had written to Russell: "Do you sometimes suspect the internal motion in M101? V.M. does a little, Hale a little more, and I much."[21]

Russell had replied, however: "I am at present inclined to believe in the reality of the internal motions, and hence to doubt the island universe theory."[22] Since Russell had been Shapley's thesis advisor and was a continuing close friend, it is ironic but possible that he inadvertently influenced Shapley in his acceptance of van Maanen's work.

Whatever the cause, by later in 1920 Shapley appears to have been completely convinced of the worth of van Maanen's measurements (perhaps, being a close friend and colleague of van Maanen, Shapley was aware of his further work on spiral rotations and became convinced by it). In a letter to Curtis, as part of their exchange of manuscripts before publication of the debate proceedings, he stated that "I have not receded at all, as you know, from my objections to the [island-universe] theory. The rotation of Messier 33 and the others (which I believe you ignore completely) will not let me back down. . . ."[23]

Of the 23 letters exchanged between Curtis and Shapley pertaining to and subsequent to the debate (from 9 June 1920 to 10 July 1921), the *only* reference to motions in spirals is the one cited above. Apparently, even though Shapley accepted van Maanen's measurements, he did not consider them to be critical.

His attitude to van Maanen's results had changed, however, from mild interest in the late 'teens to enthusiastic support by mid-1921. In answer to a letter from van Maanen[24] (which communicated the results for M81), Shapley replied:

Congratulations on the nebulous results! Between us we have put a crimp in the island universes, it seems,—you by bringing the spirals in and I by pushing the galaxy out. We are indeed clever, we are. It is certainly nice of those nebulae to have measurable motions.[25]

Russell continued to be a firm supporter and answered a 1920 letter of van Maanen's[26] with the following:

Let me congratulate you very heartily upon your detection of motion in M33. I have been waiting with great interest to see what you would get on additional spiral nebulae and this confirmation of your earlier discovery is most gratifying.

Before long I suppose that you and Peas [sic], between you will get the rotation of some nebulae with the steriocomparatore [sic] and the spectroscope too and be able to tell us just how far off it is. Then "good–bye to Island Universidus."[27]

Van Maanen's measurements of internal motions in spirals seemed to constitute strong evidence against the island-universe theory. But other findings, both observational and theoretical, were being made; although at first they appeared to support van Maanen, they soon came into direct conflict with him.

Sample Calculation:
Internal Motions in Spirals

Van Maanen's results for internal motions were strong evidence against the island-universe theory. To demonstrate this argument, consider the following:

Van Maanen's value for the mean annual rotational motion μ_{rot} of spirals is about 0."02. (He found essentially this value for each of the seven spirals measured—a remarkably consistent result.)

From the relation among proper motion (μ), rotational velocity (v), and distance to the spiral (d),

$$v(\text{km/sec}) = 4.74\mu(''/\text{yr})d(\text{pc}),$$

it can be seen that if the rotational velocity were known, the distance could be found; or, if the distance were known, a rotational velocity could be determined. Hence, if spirals are assumed to be at large distances, say 10^6pc, the rotational velocity would be

$$V_v = 4.74(0.02)\ 10^6 = 9.48 \times 10^4 \text{km/sec},$$

which would correspond to a significant percentage of the velocity of light; i.e.,

$$c = 3 \times 10^5\ km/sec,$$

$$\therefore v \approx 1/3c.$$

If a rotational velocity were assumed—a fair value to take would be on the order of 200 km/sec, as measured by Slipher and Pease—the computed distance would be

$$200 = 4.74(0.02)d$$

$$d \approx 2 \times 10^3 \text{pc}$$

This distance estimate indicated that the spirals were relatively nearby and hence small in size—certainly *not* the type of objects that would qualify as "island universes."

CHAPTER 3

Conflicts Over Spiral Nebulae

Spectrographic Problems

Van Maanen's work and the spectrographic results of Slipher, which indicated that the spirals are rotating, seemed to complement each other until a contradiction in direction of rotation was pointed out by Joel Stebbins. He wrote to Curtis early in 1924, stating that the direction of rotation found spectrographically by Slipher was opposite to that found by van Maanen.[1] Slipher's results indicated a rotation that would cause the arms to "wind up," while van Maanen's would seem to cause them to "unwind."

118

Curtis' reply revealed that he recognized the disagreement:

I had noted the apparent discrepancy you mention in your letter, but had passed it up as an error in the English, either of one or the other writer, or myself. But I have gone over again Slipher's words in comparison with one of van Maanen's photographs, and I cant make out anything else than motions in opposite directions.[2]

Slipher determined the direction by assuming that the dark lane in the nebula indicated the side nearer to the Sun. The direction of motion of the spiral arms was determined from the assumed inclination of the nebula; hence, if the dark lane were in fact on the far side of the nebula, the direction of rotation would be the reverse. In that regard, Curtis noted, "I think one may say, then, that the spectrographic results contradict van Maanen, unless one assumes that the 'lane' side is farther from us."[3]

Curtis then wrote to Slipher, asking about the apparent contradiction. He replied that one did, indeed, seem to exist.[4] Since Curtis was the chief proponent of the island-universe theory, many astronomers wrote to him asking about it. In at least one instance he answered: "The one strong argument against the theory today is found in the rotations measured by van Maanen at Mt. Wilson; these motions, though small, are too large for island universes."[5] But after receiving Slipher's letter confirming the apparent contradiction, Curtis replied to an inquiry:

As to the motions, it is true that if these are substantiated as given by van Maanen, the island universe theory must be abandoned. But the facts are, which our usually conservative English brethren seem completely to have overlooked, that the spectographic results of Slipher and others call for a motion of the spirals "like that of a spiral spring being wound up," i.e., prevailingly inward, while van Maanen's are prevailingly outward along the arms of the spiral. In this dilemma, I find myself unable to accept v.M.'s results as yet, and prefer to hold to those given by the spectrograph, as the more authentic.[6]

The problem of the direction of rotation of spirals, itself difficult enough, was further complicated by the theoretical work of Bertil Lindblad, which called for motions outward, like those found by van Maanen. These perplexing issues were not entirely settled until much later, through the definitive studies of Hubble and others.[7]

Three days after Hubble's formal announcement[8] of the discovery of Cepheids in M31 (see Chapter 4), which clearly showed the spirals to be at a great distance, Curtis wrote to Aitken:

as you know, I have always believed that the spirals are island universes, and Hubble's recent results appear to clinch this, though I myself did not need the confirmation. I have never been able to accept van Maanen's results, the main and sufficient reason being that the spectrographic results of Slipher and Pease show motion in exactly the opposite direction. And I have always been a "fundamentalist" as far as the spectrograph is concerned.[9]

Van Maanen also noticed that Slipher's results were not in agreement with his own, and he realized the contradiction much sooner than Curtis or Stebbins. In 1921, he wrote to Shapley:

I have finished M81; it is very beautiful; motion like the others, but if Slipher, in his A.A.S. note is right on the direction, my measures must be wrong! He finds motion inward, but I think he bases too much on the absorbing bands. Or, as M81 shows some increase in motion with distance from the centre, we may have the reversion of motion somewhere near the center!!![10]

The matter seems to have gone no further, however, and in fact van Maanen gained greater faith in his results as time went on, apparently disregarding the contradiction.

Slipher had become aware of van Maanen's work at least by mid-1916, when Duncan[11] wrote him about Pease's measurements of NGC 4594 and van Maanen's

James Hopwood Jeans: 1877-1946

James, the son of William Talloch Jeans (a parliamentary journalist), was born in Southport, England, on 11 September 1877. He received his early education at Merchant Taylors' School (1890 to 1896) and went to Trinity College, Cambridge, where he graduated in 1898. He was awarded an Isaac Newton Studentship in 1900 and a Trinity Fellowship in 1901, and was appointed University Lecturer in Mathematics in 1904. He was professor of applied mathematics at Princeton University, New Jersey (1905–1909), but returned to Cambridge as Stokes Lecturer from 1910 to 1912 when he resigned because of poor health. Although he held no teaching positions after 1912 he was professor of astronomy at the Royal Institution from 1935 to 1946 and a research associate of Mount Wilson Observatory from 1923 on.

In addition to his important contributions in the area of applied mathematics, such as his books: *Theoretical Mechanics* (1906), *Dynamical Theory of Gases* (1919), *Problems of Cosmology and Stellar Dynamics* (1919), *Astronomy and Cosmogony* (1928), Jeans was also a very successful popular writer. *The Universe Around Us* (1929),

Sir James Jeans, famed English theoretical astronomer and physicist. (*Proceedings of the Physical Society of London, 44*, 1932, Copyright The Institute of Physics)

preliminary work on M81 and M101. But Slipher's published and currently existing unpublished papers make no mention of van Maanen until 1924, when he replied to Curtis' letter about the difference in directions of rotation.

If Slipher was aware of van Maanen's results, as he surely must have been, it is surprising that he mentioned neither the initial confirmation of his own work nor the discrepancy in direction of rotation. A reasonable explanation is that Slipher's temperament—which kept him well out of the public view, away from scientific meetings (the 1914 A.A.S. conference was a rare exception), and his number of publications at a minimum—also kept him from publically criticizing van Maanen.

Theoretical Problems

Van Maanen also marshalled theoretical calculations that appeared to support him, but they, too, ultimately proved to be contradictory. The theorist responsible for most of this work was James Jeans. Apparently, van Maanen began considering Jeans' theory of nebular evolution[12] in 1921. By the middle of that year, van Maanen had completed measurements of internal

---

The Mysterious Universe (1930), and *Science and Music* (1938) achieved wide popular acclaim and precipitated his numerous popular lectures and BBC broadcasts.

Jeans was also well known for his love and knowledge of music—in particular, organ music—and was himself an accomplished organist. His first wife, Charlotte Tiffany, whom he married in 1907, died in 1934. After his second marriage, to Susi Hook, in 1935 (she was also an accomplished organist), the Jeans home became a center of musical, as well as scientific interest.

He won numerous awards and prizes for his theoretical research and was knighted in 1928. He became a Fellow of the Royal Astronomical Society at the age of 28, was awarded the Society's Gold Medal in 1922, and served as its president from 1925 to 1927.

Jeans was in many ways responsible for bridging the gap between classical dynamical astronomy and modern astrophysical developments. His contributions on stellar dynamics, theory of gases, quantum theory of radiation, and cosmology were among the first in the field, and were responsible for initiating further mathematical research that has changed the character of astronomy from one of almost total reliance on observations to one of observations coupled with theoretical studies.

motions in four spirals, the results of which he communicated to Shapley on two occasions. In June, he wrote that he did not think spirals possessed pure rotation and that "we may better think of Jeans' streamers from the edge of the lenticular nebulae."[13] In August, he wrote again,[14] giving Shapley a summary of his measurements; but this time he used the terms "stream" and "transverse" (Jeans' terminology) instead of the usual "rotational" and "radial." This shift was hinted at in the earlier letter to Shapley and expanded upon in a paper published in 1921: "In all cases, however, the displacements seem to represent a motion along the arms of the spiral more closely than a rotation. . . ."[15] And in the same paper, van Maanen made direct reference to Jeans' theory: "The close agreement of the direction of the displacements with the spiral arms suggests that we may have here a realization of the motions described by Jeans in his *Problems of Cosmology and Stellar Dynamics*. . . ."[16]

Jeans was aware of the relation between his work and van Maanen's as early as 1917, when he stated that their work was "entirely in agreement."[17] Also, after van Maanen's 1921 paper (in which the results for four spirals were summarized), Jeans presented the findings to the Royal Astronomical Society. In this talk, which included slides provided by van Maanen, Jeans stated that "the motion is found to be almost entirely along the arms and outward—matter is being ejected from the nucleus, and the arms show the approximate orbits of the particles."[18] In the discussion following the talk, J. H. Reynolds asked about the nature of the condensations in the arms. Jeans replied:

I see no reason to change my previous view that the condensations are masses of very rare gas—giant stars in the process of birth out of a shrinking rotating mass of gas. Van Maanen's measurements rather confirm the theory underlying this conjecture.[19]

Other theoreticians were also aware of the complementarity in the work of Jeans and van Maanen. In a review paper of 1917, A. S. Eddington mentioned van Maanen's results in relation to Jeans' work, but he remained cautious, only venturing that "it would, we think, be premature to draw any general conclusions from these observations."[20] And in a 1924 review paper on nebulae, Reynolds (a strong opponent of the island-universe theory) claimed that "in the case of the spiral nebulae, the outstanding contribution has been the measurements by Van Maanen of the motion of condensations in the spiral arms. . . ."[21]

In that paper he also discussed parallax determination for spirals and indicated that van Maanen considered a method based on Jeans' work to be one of the most reliable; then Reynolds concluded that the derived values of parallax prohibit spirals from having distances or dimensions commensurate with the island-universe hypothesis.

In late 1923, Jeans published an astonishing paper, in the hope of "provoking discussion" and "drawing attention to the importance and difficulty of the problem"[22] he was having in reconciling his calculations with van Maanen's findings. He had withheld publication for over a year, partly because of his "unsuccessful effort to find some less revolutionary interpretation of the observed motions than that here tentatively put forward. . . ."[23] He claimed a revolutionary interpretation was needed because "on the evidence of the present measurements we can only conclude that we are not free to suppose that the motions take place under gravitation, and that we must look to some new and so far unknown force for an explanation."[24] With that explanation, he then postulated

[a] generalised gravitational force which falls off approximately as $r^{-\frac{1}{2}}$ and is not directed towards the nucleus. It cannot be claimed that it has been shown such that a force exists, but the assumed existence of such a force appears to provide the simplest (and, so far as the present writer has been able to discover, the only reasonable) explanation of the motions observed by Van Maanen in the arms of the spiral nebulae.[25]

At the end of the paper he added the cautionary observation that "further measurements on nebular motions may compel their modification or abandonment,"[26] but at that time, at least, he seems to have believed them ardently: he was more willing to modify the law of gravity than to reject van Maanen.

But in 1925, shortly after Hubble's discovery of Cepheids had been announced, Jeans revised his distance measurements for spirals, employing an entirely new method based on the magnitudes of the condensations in the spirals and using Eddington's mass-luminosity law. He was then able to "dispense with v. Maanen's measurements altogether."[27] And in some notes he sent to Hubble in late 1924 (giving detailed calculation in support of Hubble's distance estimates to spirals), Jeans stated unequivocally that "van Maanen's measurements have to go."[28]

Even though Hubble's discovery of Cepheids utterly refuted van Maanen's results, the spiral rotation theory was not attacked, at least by the theorists.[29] Instead of arguing against van Maanen's results on the basis of incompatibility with their theories or contradiction with Hubble's observations, theorists merely avoided the issue. For instance, Ernest W. Brown, who like Jeans had encountered theoretical difficulties with van Maanen's results,[30] failed to refute van Maanen but rather wrote that the problem in reconciling his measurements with the law of gravitation could be avoided by taking a new approach in which "the internal motions of van Maanen play no part."[31]

Observations (van Maanen's) raised the problem for the theorists, and other observations (primarily Hubble's) resolved it.

James Jeans (*right*) and Edwin Hubble at Mount Wilson's 100-inch Hooker telescope. (*Fortune*, July, 1932)

Astrometric Problems

In addition to the spectroscopic and theoretical evidence of motions in spirals, other observational studies of proper motions also conflicted with van Maanen.

a. *Lampland.* In 1914, C. O. Lampland of the Lowell Observatory found a proper motion for NGC 4594 of $+0''.006$/yr in right ascension and $-0''.05$/yr. in declination,[32] and two years later he reported[33] proper motions and rotations for M51 and M99. His technique was to compare plates taken by Isaac Roberts in 1896 and 1898 (using a 20-inch reflector) with ones he took himself (using a 40-inch reflector); the results were $\mu_a = +0''.118$/yr and $\mu_\delta = 0''.017$/yr for M51, and $\mu_a = +0''.016$/yr and $\mu_\delta = +0''.043$/yr for M99.

124

His findings attracted little attention until he wrote another paper in 1921 on variations in the nucleus of M99.[34] Even though Lampland was uncertain of the nature of the changes and his results did not actually confirm van Maanen's work, both van Maanen and Shapley felt the findings were significant. To Shapley's question, "What do you think of Lampland and M99? Hard on the island universes, isn't it?"[35] van Maanen replied: "Of M99 we have very little news and do not know if Lampland found motions or changes in intensity, only variation. By this time Curtis and Lundmark[36] must be the only defenders of the island-universe theory."[37] But Lampland's measurements could not be verified. After studying the nucleus of M99 in 1923, Hubble wrote to Slipher (then the acting director of the Lowell Observatory) that he had found "no evidence of change this season."[38]

b. *Kostinsky.* In 1917 S. Kostinsky reported his findings of a mean proper motion for M51 of approximately $0\rlap{.}''40$/yr, but he cautioned that his error bounds were large.[39] He merely cited this finding without discussing its relation to comparable determinations by Curtis, Lampland, and van Maanen.

c. *Schouten.* An independent observation of internal motion in M51 was presented by W.J.A. Schouten in 1919.[40] While at Groningen, he found (at $0\rlap{.}''1$ from the center): $\bar{\mu}_{rot} = -0\rlap{.}''0073$/yr $\pm 0\rlap{.}''0023$ and $\bar{\mu}_{rad} = -0\rlap{.}''020$/yr $\pm 0\rlap{.}''005$; and he clearly asserted that his results were *not* in accord with van Maanen's for M101. This dissension was never mentioned in print by van Maanen. In his 1922 paper on M33, in referring to M51 he mentioned only Lampland and Kostinsky, and in his own paper on M51 he again ignored Schouten. The following year, he asserted that "similar motions [to those he had detected] have been found in M51 by Kostinsky, Lampland, and Schouten."[41] Actually, the direction of motion implied by Schouten's measurements was *opposite* to that derived from van Maanen's!

d. *Curtis.* Another dissenting view—and one of the strongest—was expressed by Curtis, who had begun studying motions of nebulae before van Maanen. Curtis never believed van Maanen's results on internal motions. He did not mention them at all in the Great Debate, perhaps because he had no clear-cut way of refuting them; moreover, in the highly edited version of the debate that was published, his only reference to them was the following:

should the results of the next quarter-century show *close agreement among different observers* to the effect that the annual motions of translation or rotation of the spirals equal or exceed 0″.01 in average value, it would seem that the island universe theory must be definitely abandoned.[42] (Curtis' italics.)

In an important lecture at that time, in which he summarized the arguments for spirals being in the Milky Way, Curtis made no mention whatsoever of motions[43]—not even his own measurements. This omission could be partially explained by Curtis' later admission to Campbell that "I have no confidence in the *real* value of either my own nebular measures or those which have been made by van Maanen."[44] (Curtis' italics.) He thought that since many of the plates were of poor quality (most had been taken with the difficult Crossley reflector), measurements based upon them could not be reliable, and remarked, "Funny; v.M. thinks I ought to measure these plates on another machine; while I have always wished he could measure some of *his* on some other machine than a big stero."[45] (Curtis' italics.)

In 1922, Van Maanen wrote to Campbell (the director of Lick), asking for permission to use the data compiled first by Curtis and later jointly by Curtis and Lundmark, and he mentioned that the information he had seen showed agreement of direction of rotation between his work and Curtis' in 75 per cent of the cases.[46] Campbell gave the requested permission, and then wrote to Curtis, telling him of the request. In that letter he candidly remarked that "I have not urged the publication of your measures, partly because of my fears that the computations were affected by computers' errors. Moore and I discovered, from looking over your measures . . . that computation is not your strong point."[47] Nevertheless, van Maanen thought they were useful, as evidenced by Lundmark's comment to Campbell that "[van Maanen] thinks that Curtis and I have got real motions for the spirals at least in the mean and he wants to see our results more in detail."[48] Even as late as 1925, van Maanen asked Curtis for permission to publish results from his data, but Curtis was adamant because he felt that "the individual measures [have not] merited publication. My present opinion is that they are worse than useless. *I very strongly prefer that you do not quote or publish these individual measures.*"[49] (Curtis'

Curtis not only doubted the worth of the proper-motion measurements, he also rejected van Maanen's findings on the basis of the spectroscopic studies, which showed the direction of rotation to be opposite that

126

found by van Maanen. Although in his private correspondence he frequently cited the spectroscopic work[50] as his reason for doubting the internal-motion measurements, he apparently did not say so publicly until 1933.[51]

e. *Lundmark.* According to van Maanen,[52] by 1921 only Curtis and Lundmark remained as strong defenders of the island-universe theory. Curtis, on the basis of the contradiction with spectroscopic evidence, rejected van Maanen's results for internal motions and parallaxes, which appeared to be evidence against the theory; Lundmark, on the other hand, rejected van Maanen's results on the basis of his own studies of nebular motions.

Lundmark came to the United States from Uppsala, Sweden, on a fellowship and spent a year at Lick (1921-1922) and another at Mount Wilson (1922-1923); hence he came into contact with both Curtis (at the former) and van Maanen (at the latter).

In a paper on M33 in late 1921,[53] about ten months after van Maanen's on the same subject, Lundmark argued for a large distance to spirals; in fact, he arrived at the same lower limit as Curtis—about 3,000 pc. Though that small distance would still leave the spirals within the Milky Way even by the standards of Kapteyn's model, as a lower limit it certainly conflicted with van Maanen's results.

In 1922, Lundmark tried another technique to demonstrate that spirals are remote. From proper motions of 23 spirals, he determined the apex of solar motion, which disagreed with the apex determined from their radial velocities. He therefore concluded that the "proper motions derived must be illusory or affected by large systematic errors,"[54] and he gave 10,000 pc. as a minimum distance to the spirals.

Although this paper was not a direct attack on van Maanen's results (it did not mention them), the conclusions of large distances and the unreliability of proper-motion data concurred with Curtis' views. The method of attack was also strikingly similar to Curtis' —proper motions were used for a large number of spirals, and the assumption was made that radial and tangential motions should be roughly equal. Since Lundmark had worked with Curtis, the similarity of approach is not unexpected.

During the early 1920s, Lundmark became van Maanen's most outspoken critic. In a comprehensive paper in 1922 on the motions of spirals,[55] he made the following important points:

Knut Lundmark. (Lund Observatory photograph)

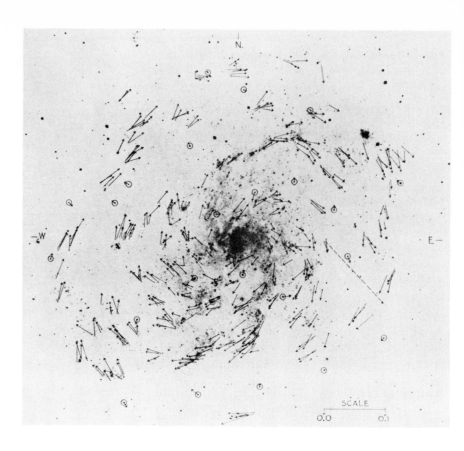

Internal motions in the spiral Messier 33 as found (*above*) by Van Maanen (*Astrophysical Journal, 57*, 1923, University of Chicago Press); (*opposite*) by Lundmark (*Astrophysical Journal, 63*, 1926, University of Chicago Press). The arrows indicate the magnitudes and directions of the mean annual motions. The comparison stars are enclosed in circles.

1 His (Lundmark's) values for the translational motions of spirals as a whole agreed with van Maanen's, but his for internal motions did not.

2 His values for internal motions yielded parallaxes about one-tenth as large as those derived from van Maanen's translational motions.

3 Van Maanen's contention that the rotational components either remain constant or decrease with distance from the center of the spiral was contradicted by Pease's spectroscopic measurements, which showed an increase with distance from the center; hence, Lundmark concluded that "it seems not unlikely that van Maanen's internal motions for spirals are measured too large at the centers."[56]

Lundmark continued the attack on van Maanen by remeasuring the plates of M33. When he completed the work in 1924 (at Uppsala) he wrote to Aitken (then the associate director of Lick):

. . . I have just finished the reduction of measures performed by me at Mount Wilson concerning the spiral Messier 33. The 400 points measured do not show any good agreement with van Maanens [sic] measures concerning the internal motions. The nebulae is so to say stationary for me and if I got a general rotational effect it is very small compared with that of van Maanens. My value for the translational motion is in excellent agreement with his.[57]

And he submitted a paper to the *Astrophysical Journal* in August 1925, whose abstract contained this statement:

The photographs measured and discussed by van Maanen . . . were measured with the same measuring instrument and discussed by the same method. The same comparison stars and nebular points were selected for measurement. Any systematic difference in results is therefore of personal origin.[58]

However, Lundmark failed to state anywhere in the paper itself that he was in substantial disagreement with van Maanen; in fact, the *only* indication of a conflict is that quoted above. If the paper is not read carefully, one might even conclude that the measurements agree. It is surprising, in the light of Lundmark's earlier statements, that he did not make the contradiction clearer. Perhaps he attempted to do so but was thwarted by his style of writing, which is difficult to decipher at best; or perhaps Shapley was correct in believing (as he stated in his reminiscences[59]) that Lundmark was too gentlemanly to criticize van Maanen publicly.

But what is even more intriguing is that Lundmark's measurements failed to receive due attention. The entirety of the published recognition by the persons involved in the controversy amounted to no more than

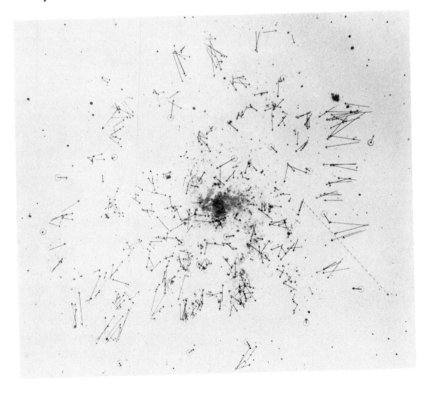

Reynolds' brief statement in a review article, and one short, noncommittal paragraph by Jeans to the effect that the results of the two men were inconsistent with each other and with Hubble's determination of nebular distances.[60] Certainly Jeans or Brown could have seized on the lack of observational verification as a way out of their theoretical difficulties; likewise, Curtis could have used Lundmark's findings to prove that he had been right all along. Moreover, these results could have been of great value to Hubble in his continuing conflict with van Maanen. In short, many individuals should have wanted van Maanen's measurements discredited, but when a real opportunity arose to do so, no one seized it.

Paradoxically, the most remarkable comments about the work came from Lundmark himself. In 1927, he published a lengthy paper (124 pages plus 11 plates), summarizing the current state-of-the-art about galaxies (which he called "anagalactic nebulae"). In the section on internal motions, he discussed in detail his measurements and reductions as well as those of van Maanen. His results naturally were the same then as in 1926; nevertheless, Lundmark concluded the section with this statement:

It is very difficult to explain the divergence between the rotational or internal motions measured by van Maanen and by me. On the other hand there is a decided correlation between the two sets of measures and there is no doubt that the phenomenon whatever it may be is *closely related to the general course of the spiral arms.* . . . Van Maanen's results may be exaggerated but may still express a real phenomenon.[61] (Lundmark's italics.)

Hence even Lundmark—who had been opposed all along to van Maanen's results, who had personally failed to verify them after years of effort, and who was undoubtedly aware that Hubble's results flatly contradicted van Maanen's—found it possible to say in 1927 that the measurements actually might be real!

Van Maanen's Response

In 1922, van Maanen attacked Lundmark's position on the island-universe theory:[62] he questioned several of Lundmark's assumptions,[63] reaffirmed his own agreement with Jeans' theory, discussed the mean parallaxes from the data of Curtis and Lundmark (stating his agreement), estimated parallaxes for spirals (from his own techniques and Jeans' theory), and concluded that the existing evidence did "not seem to warrant the acceptance of the island-universe theory."[64]

He published an even more significant paper pertaining to Lundmark in the following year. After discussing possible sources of error in the telescope, the plates, and the measuring instruments, he concluded that the motions he had detected in seven spirals must be real. And he noted that Seth Nicholson had checked and confirmed the motions he found in M101 and that Lundmark's measures of M33 agreed with his own. How he was able to make this last statement is perplexing, since Lundmark did not finish the work until he returned to Uppsala and he did not publish it until 1926. Even a private communication in 1923 could have been based on nothing more than extremely preliminary calculations.[65]

In the same paper, van Maanen again quoted the results of Curtis' study of proper motions, claiming that ". . . I have shown that Curtis' results cannot be due to the influence of accidental errors as was once assumed by Curtis himself."[66] Van Maanen cited the work of Curtis and of Lundmark in support of his own findings, although they did not concur with this claim. Either he did not understand their arguments or he did not accept them.

Also, van Maanen again called upon Jeans' work as verification of his own, quoting the following values of parallax determined from theory: $0\rlap{.}''0006$ for M31; $0\rlap{.}''0011$ for M101; and $0\rlap{.}''0065$ for M51. Then, using Curtis' mean total proper motion of $0\rlap{.}''033$/yr, together with the mean radial velocity of 600 km/sec, he found a mean parallax of $0\rlap{.}''00013$ and $0\rlap{.}''00015$, respectively. The distances corresponding to these parallaxes vary from about 150 to about 8,000 pc. Such modest distances not only required the spirals to be small in size but also doomed the island-universe concept, because estimates of the Milky Way's extent were being given from 10,000 to 100,000 pc.

In the early 1920s, van Maanen's work was widely accepted, for good reasons. He was a staff member at one of the world's most important observatories, had the finest astronomical equipment at his disposal, and he was renowned as a meticulous observer; moreover, he was supported by prominent scientists, most notably Jeans. Although evidence existed that refuted van Maanen's results, it was largely ignored. New discoveries were soon to come, however, which not only proved van Maanen wrong but also restored the island-universe theory. One man was almost single-handedly responsible for these developments: Edwin Hubble.

131

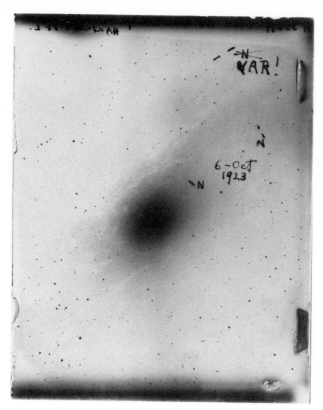

The photographic plate of Messier 31 on which Hubble discovered the first Cepheid variable star in a spiral nebula. The Cepheid is marked "VAR!" in the upper right corner and the objects marked "N" are novae.

CHAPTER 4

Cepheids in Spiral Nebulae

Period-Luminosity Relation

As shown in Section One, the development of the period-luminosity relation began in 1908 when Leavitt noticed a correlation between brightness and periods for variable stars in the Magellanic clouds. In 1913, Hertzsprung identified the light curves of these variables as being identical with the light curves of Cepheid variable stars in the Milky Way.[1] Finally, in 1916, Shapley calibrated the curve and used it to determine distances to the globular clusters that outline our Galaxy.

Hence, if one found a Cepheid anywhere in the universe, its distance could be found merely by measuring its period of pulsation (assuming, of course, that all Cepheids in the universe are the same and obey the same period-luminosity law).

Discovery of Cepheids in Spirals

In 1923, Hubble, at Mount Wilson, was able to re-solve the outer regions of two nearby spirals, M31 and M33, into "dense swarms of images which in no way differ from those of ordinary stars."[2] Several of these objects (22 in M33 and 12 in M31) exhibited the light curves of Cepheid variables. Hence, using Shapley's own period-luminosity curve, Hubble showed that the distances to the spirals were about 285,000 pc.

It should be noted here that Hubble's results did not depend on using the 100-inch telescope. He also used the 60-inch telescope (which was completed in 1908); it would have been adequate for the task. In addition, it should be noted that not only were other, larger telescopes available before 1923, but the 100-inch became operational in 1918.

Hubble based his results on three major assump-tions: 1) the variables are actually connected with the spirals; 2) there is no serious amount of absorption due to amorphous nebulosity in the spirals; 3) all the Cepheids in the universe are the same. Although the last assumption is incorrect (in the sense that the spirals are even farther away than Hubble's results show), the distances are certainly larger than Shapley believed and the spirals must be large stellar systems similar to our own Galaxy. This last conclusion is ob-vious, because for an object to have a large angular size in the sky and still be very far away, it must also have a large linear dimension. For example, for M31, the angular size is about $2°$; hence, from the relation

$$\text{angular size (radians)} = \frac{\text{diameter (pc)}}{\text{distance (pc)}}$$

the diameter of the spiral is found to be almost 10,000 parsecs.

Implications for the Nature of the Nebulae: III

Although Hubble made the discovery in 1923, he was very cautious about publishing. He wrote to Shapley early in 1924[3] announcing his discovery of the first two variables in M31, which indicated a distance of something over 300,000 pc. Shapley's reaction to this result is interesting. In his reply to Hubble he referred to the information as "the most entertaining piece of literature I have seen for a long time."[4] Ap-parently, he did not realize the importance of the discovery. In a letter to Hale he briefly mentions Hubble's discovery in one short paragraph, as if in passing: "We have just heard here of Hubble's marriage. Recently he wrote to me in some detail concerning his astonishing Cepheid variable in Andromeda."[5]

Many other astronomers, however, considered Hubble's work to be highly significant. Russell, in a letter to the managing editor of Science Service (in response to a request by the editor for a list of the outstanding advances in Astronomy in 1924), describes Hubble's discovery as "undoubtedly among the most notable scientific advances of the year."[6]

Russell's response is significant. Previously he had accepted van Maanen's measurements of internal motions in spirals, which had indicated that they must be very near, hence small in size, and certainly not "island universes" comparable to the Milky Way. Hubble's discovery, indicating that the spirals are very far away and therefore quite large in size, seems to have changed Russell's thinking.

Nor was Russell the only influential personality to change his mind because of Hubble; Jeans, who on theoretical grounds had agreed with van Maanen's smaller distances, also revised his thinking after communicating with Hubble.

Russell wrote to Hubble in late 1924,[7] offering him heartiest congratulations and urging him to announce his results formally at the upcoming meeting of the American Association for the Advancement of Science in Washington. Hubble sent a paper about his results to Russell, who read it at the meeting.

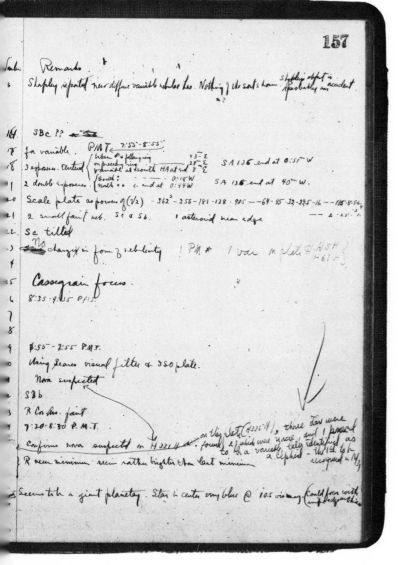

The page from Hubble's observation notebook citing the discovery of the Cepheid. See the notation next to plate number 335 on the right. (Hale Observatories photographs)

An interesting account of how Hubble's work came to be presented at the meeting is given in a letter by Stebbins:

Madison, Wisconsin
February 16, 1925

Dr. E. P. Hubble,
Mount Wilson Observatory,
Pasadena, California

My dear Hubble:

The news has finally come through *Science* that the Second A.A.A.S. prize has been divided between yourself and another worker. I hasten to extend my congratulations, something that I have been anxiously waiting to do ever since the Washington meeting.

There are several amusing and interesting incidents which led up to the recommendation of your paper which probably you would never hear of, and some of them certainly from no one but me. On the first evening of the meeting, I happened to take dinner with Russell who had arrived rather late, and one of the first things that he enquired about was whether you had sent in any contribution. On my answering no, he then said, "Well, he is an ass. With a perfectly good thousand dollars available he refuses to take it." These remarks led to some discussion, and afterwards in a group in the hotel lobby we drafted a telegram urging you to send by night letter the principal results which Russell and Shapley could make up into a paper. After this message was drafted, Russell and I started to go over to the telegraph office to send it, but on the way we stopped at the desk and put it on a regular blank. Just as we were leaving, Russell's eye caught beyond him on the floor a large envelope addressed to himself, and at the

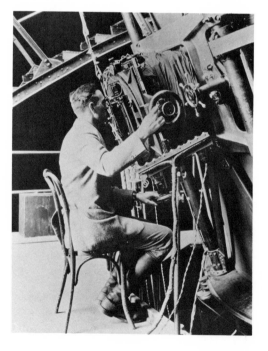

Edwin Hubble taking photographs with the 100-inch telescope at Mount Wilson in 1924. (Henry E. Huntington Library)

same time I spied your name in the upper left corner. The clerk gave us the material, and we walked back to the group in the lobby saying that we had got quick service, and that the paper was on hand. At the time, that coincidence seemed a miracle.

At the close of the meeting, the Council of the Society elected your paper as the one to be recommended for the prize, and Russell and myself were appointed a committee to see that it was properly presented to the Committee on Award. Needless to say it was Russell who drafted the report with all specifications as to the increased size of the universe, and so on, and then the material was forwarded to the committee of the Association. We learned at the time that there was going to be considerable difficulty, because a number of recommendations were coming in from the various sections.

Perhaps some of us might have been more jubilant if you had got the entire prize, but after all, you will have the satisfaction in knowing that the committee gave very long consideration to the award, and that not by mere accident has some other worthy work been left unappreciated. Not the least of my own pleasure in this matter is the recollection of the congenial visit I had with you on the mountain in one of your night watches.

With regards and best wishes, I am

Very sincerely, yours,[8]

Allan Sandage described the impact of the paper at the A.A.A.S. meeting:

The announcement of Hubble's discovery was dramatic. . . . Joel Stebbins, many years later, reminisced on this meeting and recalled that, when Hubble's paper had been read, the entire Society knew that the debate [Curtis-Shapley] had come to an end, that the island-universe concept of the distribution of matter in space had been proved, and that an era of enlightenment in cosmology had begun.[9]

Joel Stebbins himself, in a letter to the A.A.A.S. Prize Committee stated:

this paper is the product of a young man of conspicuous and recognized ability in a field which he has made peculiarly his own. It opens up depths of space previously inaccessible to investigation and gives promise of still greater advances in the near future. Meanwhile, it has already expanded one hundred fold the known volume of the material universe and has apparently settled the long-mooted question of the nature of the spirals, showing them to be gigantic agglomerations of stars almost comparable in extent with our own galaxy.[10]

Obviously, Hubble's findings had an important effect on the astronomical community. Some astronomers, however, were not completely convinced—probably because of the conflict between Hubble's results and those of van Maanen. In fact, this is what caused Hubble to remain so cautious in publishing. He said as much in a letter to Russell:

137

The real reason for my reluctance in hurrying to press was, as you may have guessed, the flat contradiction to van Maanen's rotations. The problem of reconciling the two sets of data has a certain fascination, but in spite of this I believe that the measured rotations must be abandoned.[11]

Even though, officially, Hubble's announcement was made at the A.A.A.S. meeting in December, 1924, the discovery was leaked to the press a month earlier.[12] This revelation, however, appears to have gone unnoticed.

Curtis, naturally, was pleased with Hubble's findings since they agreed with his own argument for island-universes. Writing to the Astra Klub in Yugoslavia, he stated: "I have always held this view [that spirals are separate galaxies], and the recent results by Hubble on variables in spirals seems to make the theory doubly certain."[13]

Sample Calculation:

Distances to Spirals from Cepheids

By 1924 Hubble had identified severable variable stars in M31 and M33 as Cepheids. Listed below are some of the data for the Cepheids in M31 that he published in 1925.

Log period	Maximum apparent magnitude
1.70	18.4
1.65	18.2
1.61	18.6
1.58	18.3
1.50	18.2
1.34	19.0
1.33	18.8
1.30	18.5
1.28	18.6
1.27	18.9

With Shapley's period–luminosity curve, absolute magnitude can be found for these variables. (See Section One).

For a sample calculation, take $\log p = 1.5$ and $m = 18$. Then from Shapley's curve, $M = -5$. Hence, using the relations

$$m - M = 5 \log d - 5$$
$$18 - (-5) = 5 \log d - 5$$
$$\log d = 28/5 \approx 6$$

we obtain $d \approx 10^6$ pc.

This value for the distance to M31 places the spiral well outside the Milky Way and certainly qualifies it for status as an island-universe.

A 1920 photograph of the spiral nebula NGC 1097.
(Royal Astronomical Society photograph)

CHAPTER 5

The Nature of Spiral Nebulae

Resolution of the Conflict

Hubble's distance determinations could have settled the problem abruptly in 1925, but they did not. The controversy lingered for two principal reasons: 1) Hubble's findings were based on the period–luminosity law, which was still somewhat controversial because its physical cause was unknown and its calibration depended upon the complicated procedure of statistical parallaxes; and 2) van Maanen's findings were based on direct plate measurements, which yielded remarkably consistent results. The doubts caused by the first fact were quieted within the next few years,[1] but those raised by the second were not so easily removed.

In many of his papers on the motions of spirals, van Maanen discussed possible sources of error. Indeed, in his first paper on spirals he explained how error due to magnitude effects, differences between old and new plates, and temperature effects could be avoided. He discussed these matters in more detail in the 1923 paper on M33:

That the displacements are instrumental is very improbable; there is no reason why the old plates should differ in appearance from the new ones in such a way as to produce a displacement of the nebular points with respect to the comparison stars corresponding to rotational or stream motion. It is true that in practically all cases the comparison stars were in the mean brighter than the nebular points. This might give rise to a magnitude error, but such an error could produce only a bodily shift of the nebular points with respect to the comparison stars, or a radial shift due to curvature of the field and the smaller mean distance of the nebular points from the center than of the comparison stars.[2]

In the same paper he decided that neither quality nor density of the plates could account for the displacements, and, in reference to the measuring instruments, he stated that "it is clear that defects in the optical system of the stereocomparator could never reveal themselves as a rotary motion of the nebular points, without equally affecting the comparison stars."[3]

A year later he published a paper on motions of the globular cluster M13. Because he found its internal motions to be only minute, he concluded that the results confirmed Shapley's distances and that the internal motions he had found in several spirals could not have been caused by the telescope or measuring instrument.[4]

Van Maanen continued this approach in 1927 by publishing another paper on clusters, in which he stated he had measured more of them "as a check on the displacements found in spiral nebulae."[5] He again found their internal motions to be small and concluded that errors caused by differences between plates, exposure times, and magnitudes had not affected his measurements.

(The paper also contained van Maanen's estimate for the mean parallax of the clusters—0″.000061—a value that he claimed agreed with Shapley's. Note, however, that this parallax is much smaller than any he had found for spirals, implying that spirals are considerably closer than globular clusters! It is remarkable that neither van Maanen nor anyone else seemed to have mentioned this point, since not even the most ardent opponent of the island-universe theory believed spirals to be closer than globulars.)

In an address to a special meeting of the Royal Astronomical Society in 1925,[6] van Maanen discussed the observed internal motions and possible sources of error in them. He explained that the measured displacements could not depend on the brightness of the comparison stars, since several spirals showed an increase of rotational component with decreasing magnitude, while others showed the opposite. Even as late as 1930,[7] he was convinced that his work did not contain errors, at least from effects of magnitudes, determinations of hour angles, or neglect of quadratic terms in the reductions.[8]

Thus, in the years immediately following Hubble's discovery the climate for controversy remained, because van Maanen's measurements had not been totally discredited. Not surprisingly, therefore, a few astronomers remained unconvinced by Hubble. Insight into some of the prevailing attitudes can be gained from Merrill's 1934 letter (from Mount Wilson) to Curtis:

An interesting program for our weekly Astronomy and Physics Club, on stellar and nebular distances, is being arranged. First, Adams is to outline the methods used for determining astronomical distances, then on subsequent dates Hubble and van Maanen are to follow and present the present status of the problem of spiral nebulae. They will necessarily be on opposite sides of the question, I think, as Hubble has observed quite a number of faint variables in Androm. neb. that apparently indicate a rather great distance. Van Maanen's measurements constitute about the only evidence for nearness, but they appear to represent a fact of observation, although it seems to be an almost incredible one involving the overthrow of the law of gravitation and goodness knows what else. It is, however, very hard to understand how his results can be obtained unless the apparent motions are real. The chance that it is photographic seems small and that it is instrumental still smaller. I have tried, without success, to get van Maanen to make measures to test the possibility of the results arising from obscure effects in the emulsion, but I must say that it is not very likely that such things can cause the apparent motions. Van did go as far as to measure a globular cluster in the same way as he does spirals, and found practically zero rotation with a small probable error, so this fails to support the instrumental or photographic error theory. The thing appears to me to be a real impasse.[9]

Some astronomers were equally in a quandary, as evidenced by Aitken's 1925 letter (from Lick) to Curtis: "Hubble's recent discoveries are strong arguments on your side of the question though I am not yet prepared to say that they clinch the argument."[10]

But many astronomers, probably the majority, were convinced immediately. Russell, for instance, recognized the importance of Hubble's discovery at first sight. On 12 December 1924, he wrote to Hubble and offered congratulations,[11] the same day that he wrote to the managing editor of Science Service.[12] This position was a complete turnabout from his view of only ten months earlier, when he had stated in lectures at Toronto[13] his certainty that van Maanen's evidence was correct and that spirals are nearby. Hence, for Russell at least, Hubble's discovery was crucial and decisive.

Curtis, however, did not state his opinion about the controversy in print until 1933; yet even then he failed to appeal to Hubble directly:

> The measures of van Maanen and the conception of the spirals as individual galaxies can not both be true, unless we are willing to assume velocities in the spiral arms which must occasionally amount to one-third the velocity of light. It seems impossible, moreover, to reconcile these values with the direction of the spectrographic velocity of rotation.
>
> The intervals available for van Maanen's measures were under twenty years. There seems at present no escape from the conclusion that these carefully made measures are subject to some instrumental error as yet undetected, and that

---

Adriaan van Maanen: 1884-1946

The descendant of a long line of aristocrats, Adriaan was the son of John Willem Gerbrand and Catharina Adriana (Visser) van Maanen. He received his B.A. (1906), M.A. (1909), and Sc.D. (1911) from the University of Utrecht. From 1908 to 1911, he worked at the University of Groningen, where he met J. C. Kapteyn. In 1911, van Maanen joined Yerkes Observatory as a volunteer assistant; in 1912, with Kapteyn's recommendation, he was appointed to the staff of the Mount Wilson Observatory. In his job there, measuring the proper motions and parallaxes of stars, he employed the skills he had acquired while working on his thesis, "The Proper Motions of the 1418 Stars in and near the Clusters h and X Persei."

Van Maanen used the 60–inch telescope at the 80–foot Cassegrain focus for parallax determinations; that was the first use of a reflector for such delicate measurements. In the 1920s he also began using the 100–inch reflector at the 42–foot Newtonian focus. To measure either parallax or proper motion from a photograph, he employed a stereocomparator. After sets of comparison stars were brought into coincidence, the distance separating the two images in question was measured with a movable micrometer thread.

He studied the proper motions of planetary nebulae, globular and open clusters, faint stars in or near the Orion Nebula, nearby bright stars with large proper motions, faint stars in 42 of Kapteyn's Selected Areas, and spiral nebulae.

they must be rejected until confirmation is secured by other observers and with materially increased time intervals. In view of the numerous other lines of evidence which now point so unequivocally to the adequacy of the island universe theory of spirals, no other course is open.[14]

Following Hubble's discovery, van Maanen was no longer actively supported or rejected; rather, his work merely faded into obscurity. After his 1923 paper, he published nothing more on the subject until 1930, when he released Mount Wilson Contributions No. 407 and 408. There was a revealing change in the titles of these papers, however; even though they were continuations of his lengthy series on "Investigations on Proper Motions" (fifteenth and sixteenth papers), their subtitles were altered from "Internal Motion" to "Proper Motion." Nevertheless, at the end of No. 407 he estimated the parallax of NGC 4051 from his value of its proper motion, together with its radial velocity of 650 km/sec; then he stated:

if, on the other hand, we apply Hubble's results to N.G.C. 4051, we should have to admit a distance about one hun-

Adriaan van Maanen. (Courtesy of the *Publications* of the Astronomical Society of the Pacific, *58*, 1946)

In 1916 he published his results, taken from the displacements of 87 nebular points on two pairs of plates of M101. He detected a rotation rate of 0."02/yr at a distance 5' from the nucleus. In the famous Shapley-Curtis debate in 1920, Shapley, van Maanen's life-long friend, cited van Maanens' findings as proof that spirals are relatively nearby. Fifteen years later, after plates of several spirals had been taken at longer intervals, Edwin Hubble and van Maanen published papers in the *Astrophysical Journal* stating that van Maanen's results on rotations had been incorrect, apparently because of systematic errors.

Van Maanen also attempted to measure the general solar magnetic field, an undertaking begun in 1908 by G. E. Hale.

Van Maanen's entire career dealt with measuring visually almost imperceptible changes on photographic plates. In parallax observations, he ranked highly; however, his results on the proper motions of spirals and the magnetic field of the Sun were largely incorrect. Numerous explanations have been offered as the reason why he made such errors. Whatever the correct answer may be, the fundamental fact remains that the changes he was attempting to measure were at the very limits of precision of his equipment and techniques.

A region of the spiral Messier 31 showing a Cepheid (center) and several star clusters (Hale Observatories photograph)

dred times greater than that corresponding to the foregoing parallax, and the motion derived here would then represent a velocity of the order of 65,000 km/sec., a quantity which is difficult to accept.[15]

Between 1925 and 1929, Hubble published three exhaustive papers[16] dealing with spirals as island universes and demonstrated that they are at enormous distances—ranging from 240,000 to 275,000 pc. No reference was made, however, to the conflict with van Maanen; in fact, he was mentioned only once, in a footnote.

In essence, after 1925 van Maanen's work on internal motions was largely ignored. There was no great controversy in the literature, as in the case of the debate about the island-universe hypothesis during the early 1920s. Van Maanen's work during this period was almost completely confined to proper motions of stars and planetary nebulae. Except for the two papers of 1930, which dealt with possible sources of error, he never again published research on spirals.

The public coup de grace came in 1935, when Hubble published in summary the results of measurements he had made, with Walter Baade and Seth Nicholson, of M33, M51, M81, and M101.[17] Even though they had used many of the same plates as had van Maanen, they had been unable to find the displacements; moreover, since with other pairs of plates the time interval between old and new plates was by then longer, the displacements, if real, should have been even larger and thus more easily detectable. Hubble concluded that his negative result proved that systematic errors were "obviously present" in van Maanen's original measurements.

Van Maanen also remeasured M33 and M107 (and added M74) in 1935, but he then found the displacements to be only about half the size of those he had detected in the early 1920s.[18] Although he still found measurable displacements, whereas Hubble did not, he admitted that "in consideration of the difficulty of avoiding systematic errors in this special problem, these results, together with the measures of Hubble, Baade, and Nicholson which are given in the preceding article make it desirable to view the motions with reserve."[19]

Thus, the last stumbling block to the full acceptance of the island-universe theory was removed—the conflict between the two opposing sets of evidence, Hubble's and van Maanen's, was finally resolved.

It is clear that van Maanen's initial motive for studying spirals was to search for internal motions, *not* to estimate their distances. Since virtually all of his research before 1916 had been on proper motions of stars, that he turned to searching for motions in spirals is not surprising. He had established himself as a capable measurer of small displacements on photographic plates; to measure them for spirals was merely a continuation of this type of analysis.

At first he did not seek evidence bearing on the island-universe theory, since it had not yet become a major issue; it became one shortly thereafter, however, when novae were discovered in spirals (principally by Curtis and Ritchey) and a new estimate was made

---

TABLE 5.1 Comparison of the rotational components (μ_{rot}) of internal motion in M33 found by different observers.

Observer	Date	$\mu_{rot}["/\text{yr}]$[*]	Notes
van Maanen	1921	$+0.020 \pm 0.003$	at 25-ft focus
		$+0.014 \pm 0.004$	at 80-ft focus
van Maanen	1923	$+0.020 \pm 0.001$	
Lundmark	1923	$+0.0016 \pm 0.0065$	published in 1926
van Maanen	1935	$+0.013$	at 25-ft focus
		$+0.009$	at 80-ft focus (no error bounds given)
Hubble	1935	-0.0001 ± 0.0023	longest interval
		-0.0000 ± 0.0024	mean for 2 pairs of plates

[*] Positive μ_{rot} indicates motion outward along the spiral arms.

of the extent of the Milky Way (principally by Shapley). At the outset, at least, van Maanen was not at the heart of the ensuing controversy: Curtis started it, Shapley responded by using van Maanen's findings, and suddenly van Maanen was deeply involved.

Van Maanen's results for spirals indicated that they could not be at vast distances; thus he became the champion of those who opposed island universes. Although he did not initially seek this role, after 1921 he accepted it enthusiastically. He was extremely effective in it, for as J. D. Fernie has noted, his results "were the prime evidence against the island universe theory, and they probably swayed a good many astronomers who otherwise might have inclined towards it."[20]

During the early 1920s, van Maanen's results *were* believed despite accumulating evidence to the contrary—Slipher's contradiction in direction of rotation, Jeans' difficulty in reconciling the results with gravitational theory, and Lundmark's failure to reproduce the results using the same techniques. It was not until Hubble's discovery of Cepheids in spirals that van Maanen's results began to be questioned seriously. Although the Cepheids demonstrated conclusively that spirals are remote, not everyone abandoned van Maanen's measurements immediately. Since his results were thought to represent observational facts, they were difficult to discard, even in the face of contradictory findings.

The construction of the 100-inch telescope at Mount Wilson: Work on the telescope mounting. (Hale Observatories photograph)

The 100-inch mirror of the telescope. (*Popular Astronomy*, 61, 1967)

After 1925, however, van Maanen's work was not so influential as before. The issue was almost forgotten until resurrected by Hubble in 1935, when he decided to settle the matter by measuring plates himself. Hubble's negative results in this study forced van Maanen to admit the real possibility of systematic errors in his results.

The demise of van Maanen's findings did not come easily. He continuously sought verification and corroboration for his work, sometimes going too far in this quest. For instance, he used data from Curtis and Lundmark even though Curtis warned him that they were unreliable; in addition, he claimed agreement with the results of Schouten and Lundmark even though no agreement existed. But in fairness to van Maanen, it should be noted that in his research he continuously worried about errors and in his papers he frequently discussed possible sources of them.

Van Maanen's findings ultimately were ignored not because they were directly proved incorrect but rather because the new evidence favoring spirals as galaxies was incontrovertible. Any negative proof is always difficult to make. In this case, astronomers were faced with specific observational findings from van Maanen. No appeal to theory or intuition, or even to other observations, could rightly invalidate them. But in time so much contradictory evidence accumulated that the presence of error in van Maanen's work was inescapable and hence the only recourse was simply to ignore his findings.

A near disaster for the base of the telescope tube while being brought up the mountain. (*Monthly Evening Sky Map*, June 1917)

In many instances in science it would be irresponsible of scientists to ignore evidence from a single observer that was contrary to contemporary beliefs; some of the major advances in science have come from such observations. And in the early 1920s that was, in fact, the type of finding that van Maanen's appeared to be. But then other astronomers not only accumulated contradictory evidence but also repeated van Maanen's own measurements, finding far different results. Thus, the astronomical community rightly began to ignore van Maanen—not because he contradicted their views but because his work could not be confirmed.

Source of van Maanen's Error

Although it was appropriate for the scientists at the time not to seek an explanation for the source of van Maanen's error, it is within the province of the modern historian to do so.

By 1935 it was definitely known that van Maanen's results were wrong because of a systematic error; however, its exact nature was not specified, and in fact is not known even today. An explanation has been offered by Baade[21] that on Ritchey's plates the geometrical center of the images did not coincide with the center of emulsion density, so that a systematic displacement occurred, depending on the magnification used in measuring. Such an effect, however, would not always lead to apparent motions along the spiral arms or to consistent rotation periods. The remarkable consistency of van Maanen's results is one of their most outstanding features, making them difficult to explain by systematic errors. His results for μ_{rot}, and hence for rotation period, were almost identical for every spiral.

148

It is hard to imagine an error that could yield such consistency. In particular, errors due to the plates would be unlikely to produce a displacement in the *same* direction by nearly the same amount for *every* spiral.

Baade has also suggested that residual coma could have produced the reported rotation. This explanation is also inadequate in the face of van Maanen's consistent values; moreover, a recent historian has questioned why the later measurements of Hubble (and Baade) were not influenced by the same effect.[22]

Shapley[23] and others[24] have suggested that possibly van Maanen found what he wanted to find. Although it is distinctly possible that he was unconsciously biased, inadvertently finding what he needed, there is no evisting evidence that he did so intentionally.

A new analysis has recently been performed in an attempt to describe a reasonable way in which an error such as van Maanen's could have occurred.[25] The following were considered as possible sources of error:

The completed Hooker telescope (Hale Observatories photograph)

1) Instrumental sources
   a) errors due to telescopes
   b) errors due to plates
   c) errors due to measuring instruments
2) Computational sources
   a) errors due to invalid procedure
   b) errors due to reduction formulae
   c) errors due to computational blunders
   d) errors due to poor assumptions
3) Personal sources
   a) errors due to identification of points on plates
   b) errors due to choice of objects measured
   c) errors due to interpretation of results
   d) errors due to mistakes in determining position
   e) errors due to bias in measuring

Instrumental sources can be excluded from further consideration, on the basis of the consistencies of van Maanen's results and their inconsistencies with those of Lundmark and Hubble. That is, it is unlikely that instrumental errors could yield the consistency found by van Maanen; and instrumental errors, if present, should also have affected the results of Lundmark and Hubble in the same way. That they did not is evidence of their absence.

Computational sources can also be excluded on the basis of the results of a thorough computer analysis of van Maanen's original raw data. Although a few inconsistencies were discovered, no systematic error was found.

Thus, since only personal sources remain, it is reasonable to conclude that such an error was, indeed, responsible for van Maanen's spurious internal motions.

The matter can be pursued further by attempting to describe a way in which such an error could actually occur in practice. This is a significant point to investigate, since personal errors are extremely dangerous in science and extremely difficult to identify. An identification of this error could be very important in aiding other scientists in their investigations because once forewarned, they can take steps to avoid it.

A further computer analysis was performed in which various types of errors in measurement were assumed and incorporated into the computational scheme actually employed by van Maanen. The analysis showed that a systematic error of only 0.002 mm in measuring the positions of points on photographic plates (*i.e.*, on the order of the accuracy obtainable with the measuring engines) could produce the reported internal motions, provided the direction of the measurement error is such that it is consistent with the spiral features of the plate. That is, if van Maanen

150

had a slight personal bias toward believing that the spirals were in rotation (a bias easily created merely by looking at a picture of a spiral), his results would reflect this bias. This could account for the remarkable consistency of his results because the measurements made at the very limit of perceptibility are extremely sensitive to precisely such errors.

The Direction of Rotation

The 1935 papers of Hubble and van Maanen, published back-to-back in the *Astrophysical Journal*, finally brought to an end this aspect of the controversy over distances to spirals; but the question of their actual direction of rotation lingered for another eight years, because even in 1935 van Maanen claimed that "although my own measures of recent plates . . . show considerably smaller values of the apparent rotational component than those first obtained, the persistence of the positive sign is very marked and will require the most searching investigation in the future."[26] (The positive sign indicated that the arms are "leading"; *i.e.*, spirals rotate so as to unwind their arms.)

The question of direction of rotation was not new: Slipher had spectrographically found the arms to be trailing, while Lindblad[27] had theoretically argued that they were leading.

Night view of the valley below Mount Wilson, showing the lights of Los Angeles, Hollywood, and more than forty other cities. (Hale Observatories photograph)

The argument centered on a prominent feature in many spirals—the "dark lane"—which was generally believed to be caused by absorption.[28] The critical problem was to determine, for a given spiral, on which side the observed lane was located. The spectrograph gave directly the sense of rotation ("left-handed" or "right-handed"). If the lane was located on the near side of a "right-handed" spiral (and due to obscuring matter in that half of it), then the arms must be trailing; conversely, if the lane was located on the far side of a "right-handed" spiral (and due to obscuring across the entire object), then the arms must be leading.

After thoroughly investigating the problem, in 1941 Hubble sent Slipher a summary of his conclusions:

For a number of years I have been discussing the interpretation of absorption patterns in relation to tilt, with every group I met—and have found a prevailing skepticism on all interpretations other than those cases in which a heavy peripheral band is silhouetted against a nuclear mass (in none of which could the spiral pattern be traced). Here, for instance, Baade considered the direction of rotation as not demonstrated, and, on theoretical grounds, was inclined to favor Lindblad's view. The others simply said no definite conclusions were possible on the basis of then available data. I discussed the matter at length at the meeting at the MacDonald Observatory, at a special conference with Oort, Lindblad, Mayall, and our local men, and with various other astronomers (even the Harvard people). The outcome was a very decided opinion that the direction of rotation was a fundamental datum for dynamical studies but that the question was as yet undetermined.

It was for this reason that a special search for unambiguous cases was made by examining every one of the Shapley-Ames nebulae in the northern sky. I believe the most favorable object is the recently observed NGC 4216. Your original interpretation was correct but, until the new objects were observed for rotation, it was considered as one of two possible interpretations, each of them supported by plausible arguments. This position, let me emphasize, was taken by practically all the men I consulted, observers as well as theoretical men.

Well, at any rate, there can be no uncertainty now, and the theoretical men can get on with the theory of spiral structure. That is the main thing.[29]

Hubble finally published the results two years later,[30] giving a strong argument that the direction of rotation must be such that the arms are trailing. Thus, even van Maanen's results for the direction of the rotation were shown to be wrong.

There can be no doubt now that van Maanen's results delayed acceptance of the island-universe theory. Shapley became an ardent believer in them,

Bertil Lindblad at the Astronomical Observatory of the University of Upsala, Sweden, in 1925. (Yerkes Observatory photograph)

using them in his arguments against island universes. Jeans was so affected by them that he even proposed modifying the law of gravitation so that theory would fit observations. And Hubble, whose distances to spirals disagreed with the implications of van Maanen's findings, withheld announcement of his discovery for a year because of the contradiction.[31]

For over a decade, van Maanen's specious results confused the island-universe issue and impeded research in extragalactic astronomy. His problem arose from working at the very limit of contemporary science and technology. Today, cosmologists and other astronomers, by necessity, continue to press their equipment and techniques to the limit. Thus, the lesson of van Maanen should not be forgotten.

A 1920 photograph of the spiral Messier 74 taken with the 60-inch telescope at Mount Wilson. (Royal Astronomical Society photograph)

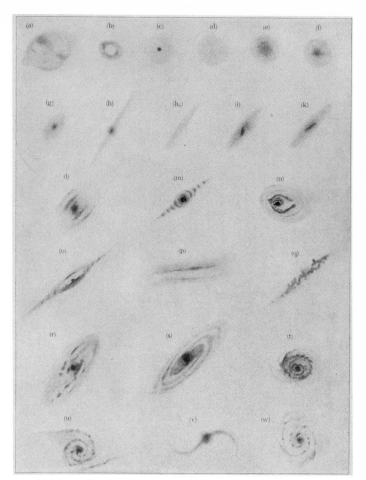

The set of drawings used by Max Wolf to classify nebulae. (Reproduced from Hubble's Ph.D. dissertation, "Photographic Investigations of Faint Nebulae," University of Chicago, 1917)

CHAPTER 6

The Evolution of Nebulae

Classification of Nebulae

One of the questions raised by the discovery that spiral nebulae (and other types of nebulae) are external galaxies is, what is the structure and evolution, if any, of these objects? Hubble first turned to this question in 1922, before he had shown that they were external galaxies.

Even though the nature of nebulae remained unknown until fairly recently, their classification according to their observable characteristics was possible long ago, as the early schemes of William Herschel[1] and Max Wolf[2] attest. Herschel's cumbersome system (Table 6.1), consisting of a set of capital and lower case letters used to describe the appearance of a neb-

154

TABLE 6.1	W. Herschel Classification System
B. Bright	v. very
F. Faint	c. considerable
L. Large	p. pretty
S. Small	e. extremely

ula, involved many characters and required a key. For example, vBpLvgmbM meant "very bright, pretty large, very gradual, much brighter in the middle."

Herschel's son John introduced a system of five categories, each with five subdivisions, as shown in Table 6.2. An object could then be described by using five numbers. For example, 32255 would mean "middle-sized, bright, round, discoid, milky." Obviously, this system is not much of an improvement.

Wolf used 23 drawings of nebulae as standards, each being assigned a small-case letter for identification; this system also needed a key.

In 1917, Hubble summarized the state of knowledge about nebulae in a paper based on his dissertation: "Extremley little is known of the nature of nebulae, and no significant classification has yet been suggested, not even a precise definition has been formulated."[3] In this article he used the classification scheme of Wolf (although he added an intermediate class, g_0).

Feeling that the classification schemes of Herschel and Wolf were not adequate, Hubble set about developing one of his own. In 1922, he suggested a scheme based upon the "fundamental differences between galactic and non-galactic nebulae."[4] These terms carried no implication about the nature of the nebulae, but described the observed position: "non-galactic" meant that members of their class tend to "avoid the galactic plane and to concentrate in high

TABLE 6.2   J. Herschel Classification System

Subclass	Magnitude	Brightness	Roundness	Condensation	Resolvability
1	Great	Lucid	Circular	Stellate	Discrete
2	Large	Bright	Round	Nuclear	Resolvable
3	Middle-sized	Faint	Oval	Concentrated	Granulate
4	Small	Dim	Elongate	Graduating	Mottled
5	Minute	Obscure	Linear	Discoid	Milky

155

galactic latitudes."[5] Hubble also discussed the view expressed by Curtis in 1919 that all nebulae could be divided into three groups—planetaries, diffuse, and spirals. He thought the classes of "planetary" and "diffuse" were adequate to describe galactic nebulae, but non-galactic nebulae required a more complete description than merely "spiral." As he put it, his scheme of classification was "a compromise between Curtis' generalization and Wolf's specialization."[6]

Apparently, Hubble's scheme had its origin in a report prepared for the first meeting of the International Astronomical Union (I.A.U.), which had been formed at Brussels in 1919. In May 1922, the I.A.U. held its first meeting, whose purpose was to summarize contemporary research, to make resolutions toward greater uniformity in nomenclature, and to promote international cooperation among astronomers.[7] The I.A.U. established commissions on various aspects of astronomical study. It was in the report of the American Section of the Commission on Nebulae that Hubble first proposed his classification:

> There are a number of facts connected with nebulae that will have more or less significance in their classification, although the full relative importance of these is not well enough known to attempt to classify the nebulae on a rational physical basis. A number of descriptive terms are in fairly common use (spiral, spindle, irregular, diffuse, planetary, stellar, galactic, gaseous, white, self–luminous, dark and reflection), and as work goes on, a simple descriptive classification, using these familiar terms may be useful.

Hubble as a graduate student at Yerkes Observatory, 1914 (*top row, third from left*). (Yerkes Observatory photograph)

GALACTIC	NON-GALACTIC
1. Planetary	1. Spiral
2. Diffuse	2. Spindle
a. Luminous	3. Ovate
b. Dark	4. Globular
	5. Irregular

Galactic nebulae classify readily in a fashion that permits sub-divisions by special investigators. Non–galactic nebulae are otherwise, and the sub-divisions here suggested are very tentative. They are a middle course between Curtis' generalization and Wolf's specialization.[8]

Early photograph of Hubble: Oxford University portrait, *c.* 1914. (Henry E. Huntington Library)

One of the recommendations of the Commission on Nebulae was the necessity of a homogeneous system of classifying and describing nebulae.[9] The report of the commission, prepared by its president, G. Bigourdan, proposed a classification scheme similar to that of John Herschel.[10] Bigourdan's report made no reference to the recommendations of the American Section, with the result that Hubble's comments on a classification scheme were omitted from the official transactions. Knut Lundmark commented on this matter in a letter to W. W. Campbell:

> The report on nebulae was evidently written by Bigourdan. I admire Mr. Bigourdan's work but I must repeat that I felt disappointed at reading his report.
> Dr. Slipher had excellent suggestions and Mr. Hubble had written an extensive general program, which was sent to the committee and as far as I can judge nothing of that is taken up.[11]

Apparently Bigourdan chose to announce only his own published scheme of classification, rather than that proposed by the newcomer Hubble.

Although Hubble received his Ph.D. from the University of Chicago in 1917, he did not begin his professional career until 1919 when he joined the staff at Mount Wilson. The reason for the delay was World War I. After staying up all night to finish his Ph.D. thesis, and making his oral defense the following morning, he joined the army.[12]

Hubble's early ideas on nebular classification were based entirely on observational evidence; as he put it, the scheme "must rest solely upon the simple inspection of photographic images."[13] Nevertheless, the theoretical prospects of an evolutionary sequence also appealed to him, as is clear from the American Section Report: "We seem to be succeeding with the evolutional sequence classification of the stars, and we may look forward with some hope to a time when something of the sort can be attempted with the nebulae."[14]

157

Captain Hubble with his sister, Lucy, *c.* 1917. (Henry E. Huntington Library)

Arguments from Theory

In 1923, Hubble wrote to V. M. Slipher, who was president of the I.A.U. Commission on Nebulae and hence the center for all communications concerning nebulae, giving more definite ideas about a criterion to be used for the classification scheme—the nebular evolution theory of Jeans.[15] Although Hubble was considering the evolutionary theory, he acknowledged that classification should be based on observed characteristics. In a letter to Slipher he stated:

I have been trying to construct a classification of non-galactic nebulae analogous to Jeans' evolution sequence but from purely observational material. The basis is a distinction between amorphous nebulosity and the granular beaded arms of spirals. As a first approximation the following suggests itself.

Amorphous	A0	such as M87 or N.G.C. 3379
	A1	M32
	A2	M59
	A3	N.G.C. 3115
Spirals	S0	N.G.C. 4594
	S1	M81
	S2	N.G.C. 2841
	S3	M101

In addition are the occasional ϕ shaped spirals (M95), S shaped spirals, and irregular nebulae such as M82, N.G.C. 4214, etc.[16]

158

Later in the letter he continued:

The agreement in mean total magnitudes suggest [sic] that the quantities of material in the types are of the same order and hence it is quite possible to conceive of them as representing different stages of an evolutionary sequence.[17]

The Jeans theory that one type of nebula evolves into another was generally accepted by H. Jeffreys, among others, as is evidenced by a 1923 paper:[18]

The predications of this theory [Jeans' theory of the origin of spiral nebulae] are so strikingly in accordance with the results of modern observations of the forms and motions of lenticular and spiral nebulae, as to afford good evidence for the truth of this theory or some closely similar one. . . .

In the conclusion of the article he remarked that

the forms of lenticular nebulae agree with the forms predicted by Jeans' theory. . . . The spiral nebulae, to judge from the continuous gradation that appears to exist between the lenticular and spiral forms, are probably a later stage in the evolution of the lenticular nebulae, starting from the stage when the lenticular nebulae develop sharp edges.

The classification scheme proposed by Hubble in the letter to Slipher differed considerably from the one published in 1922. The categories of ovate, spiral, and globular no longer appeared. He apparently wanted to separate non-galactic nebulae into only two classes, with intermediate gradations as suggested by Jeans' theory. Hubble admitted, however, that this scheme was merely his first arrangement. Later in 1923, in another letter to Slipher,[19] Hubble expanded the idea and strengthened his arguments. Along with the letter he included a manuscript describing his plan for the classification of all nebulae on the basis of photographic images. The idea of an evolutionary sequence was obviously on Hubble's mind, since he stated that "the observer may well look to Jeans' theory for the thread of physical significance that shall vitalize a system of classification of non-galactic nebulae."[20]

Hubble remained cautious, nevertheless, and stated further that "in the scheme presently to be proposed, a conscious attempt was made to ignore the theory of Jeans and arrange the data purely from an observational point of view."[21]

This statement was essentially the same as the one he had made in the earlier letter to Slipher; however, the classification proposed in the manuscript was entirely different. The scheme for non-galactic nebulae was given as follows:[22]

Classes	Symbol	Examples	
A. Elliptical	En		
		3379	E0
n=eccentricity[23]		4821 (M32)	E2
		4621 (M59)	E4
		3115	E7
B. Spirals			
1. Logarithmic[24]	S		
a. Early	Sa	4594	
b. Middle	Sb	2841	
c. Late	Sc	5457 (M101)	
2. Barred Spirals[25]	SB		
a. Early	SBa	4754	
b. Middle	SBb	3351	
c. Late	SBc	7479	
C. Irregular	I	2336, 4449	

In this ordering the spirals were listed as a group different from the ellipticals, as in the earlier 1923 scheme. Note that he replaced "amorphous" with "elliptical" and listed them first in the classification

Walter Adams, James Jeans, and Edwin Hubble (*l. to r.*) at the Mount Wilson Observatory. (O. Struve and V. Zebergs, *Astronomy of the 20th Century*), Macmillan: New York, 1962)

because, in his words, "there is some justification in considering the elliptical nebulae as representing an earlier stage of evolution. . . ."[26]

The use of the terms "early," "middle," and "late" in this classification was crucial; apparently Hubble was firmly convinced that an evolutionary sequence exists, and he used words with a temporal implication to suggest it. In fact, in discussing spiral nebulae, he stated:

Two extreme types are conspicuous. The one has a relatively large amount of amorphous nebulosity immediately about a bright condensed nucleus. From this central region spring closely coiled spiral arms in which are little or no granulation. . . . The other type has a fainter almost stellar nucleus surrounded by very little amorphous nebulosity. The spiral arms are much closer to the nucleus, are more open—unwound as it were—and are conspicuously granulated . . . the gap between the two extreme forms is well filled and a series is readily constructed in which the granular spiral arms seemingly grow at the expense of the amorphous region, unwinding as they grow.
Carrying the series from the second stage toward the first, the increasingly amorphous central region and the diminishing relative importance of the spiral arms logically lead to the thin discus shaped and wholly amorphous elliptical nebulae. Likewise assuming the existence of the series among the elliptical nebulae ranging from globular to discus shaped, an extrapolation made easy by Jeans' theoretical discussions, leads to the first form of spiral. There is thus some grounds for using the terms early type and late type spirals and considering the elliptical nebulae and spirals as a single evolutional sequence.[27]

Even though he seemed to be sure of his conclusions, he did not publish the scheme. In a letter to Slipher he explained that instead of publishing it at that time he wanted to submit it to the members of the Committee on Nebulae for their comments, "the outcome of which might be a system of classification approved by the committee and sanctioned by the I.A.U."[28] Again, in early 1924, Hubble wrote to Slipher that "Mr. Hale thinks I should publish the system of classification. I would prefer that it go through the committee if that is feasible within a reasonable short time."[29] Hubble wanted confirmation of his ideas before publishing a scheme that rested upon Jeans' speculative theory of nebular evolution.

Finally, two years later, Hubble found what he needed. In a letter to Slipher he announced the justification he had been seeking:

As for the classification, I have found a complete justification for the proposed sequence as a basis, in the analysis of Holetschek's nebulae.[30] The results are being typed now and I will send you a manuscript. The essense is in the relation

$$M_T = C - 5 \log d$$

where M_T and d are the total mag. and the diameters.[31] C increases smoothly from the globular nebulae to the late type spirals, Sc. The slope is that of the inverse square law! C measures the expansion throughout the sequence for nebulae of a given luminosity.[32]

He finally published[33] the scheme almost exactly as it had appeared in the 1923 manuscript.[34] This arrangement, with a few later modifications by Hubble, is now famous as the "tuning fork" diagram.[35]

Hubble remained cautious when he published his classification scheme in 1926. Although he stated that nebulae fall into progressive sequence and that "the various stages in the sequence represent different

Spiral galaxies classified according to Hubble type. (Hale Observatories photograph)

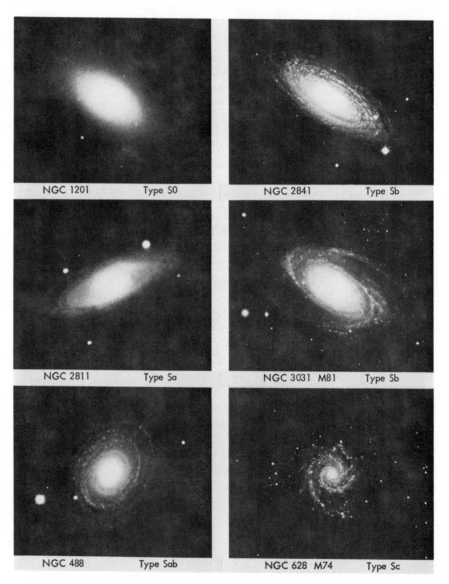

NGC 1201 — Type S0

NGC 2841 — Type Sb

NGC 2811 — Type Sa

NGC 3031 M81 — Type Sb

NGC 488 — Type Sab

NGC 628 M74 — Type Sc

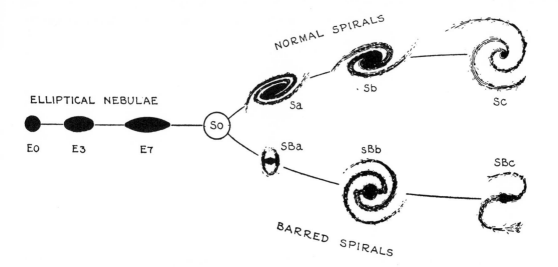

ELLIPTICAL NEBULAE

EO   E3   E7   So   NORMAL SPIRALS   Sa   Sb   Sc

SBa   sBb   SBc   BARRED SPIRALS

phases of a single fundamental type,"[36] he did not claim to have discovered observational proof of Jeans' evolutionary theory. In fact, he wrote that

The classification sequence of galaxies developed by Hubble. (E.P. Hubble, *Realm of the Nebulae*, Yale University Press: New Haven, 1936)

although deliberate effort was made to find a descriptive classification which should be entirely independent of theoretical considerations, the results are almost identical with the path of development derived by Jeans from purely theoretical investigations. The agreement is very suggestive in view of the wide field covered by the data, and Jeans' theory might have been used both to interpret the observations and guide research. It should be born in mind, however, that the basis of the classification is descriptive and entirely independent of any theory.[37]

Perhaps Hubble's reluctance to hold strongly to nebular evolution stemmed partly from the cool reception his scheme had received at the Cambridge meeting of the I.A.U. in 1925. The I.A.U. Committee on Nebulae and Clusters[38] decided that

such systems as that recently proposed by Dr. Hubble, which employed terms suggestive of certain physical properties of the nebulae about which there was still much doubt, should be avoided, and that a simpler system of a more purely descriptive nature should be used. . . .[39]

Since Hubble had refrained from publishing the scheme he developed in 1923 so that the I.A.U. could approve it, the rejection might have led him to take a more cautious position. Even though he found apparent justification for the scheme in 1926, he remained conservative in publishing. In addition, of course, being a good scientist, he was usually cautious in his conclusions. In fact, N. U. Mayall, who wrote his thesis under Hubble, has described him as an extraordinarily

163

exact and careful scientist who normally refrained from assertions that were not well supported by evidence.[40] Even so, Hubble's caution in publishing about nebular evolution is remarkable, considering his comments of 1923.

Jeans was naturally eager to accept Hubble's results since they fitted his theory almost exactly. In a 1931 publication, he stated:

I arrived at a sequence of shapes which agreed, almost exactly, with that which Dr. Hubble subsequently found. . . . This leaves little room for doubt that the nebulae we see in the sky are members of the theoretical sequence, that they began as rotating masses of gas, and that we see them in various stages of development.[41]

As might be expected, an evolutionary ordering had also occurred to other investigators; for instance, in 1924, H. N. Russell remarked that ". . . Jeans' theory seems to be matched by the things we have in the sky. . . . It looks as if we had stages of evolution shown."[42] It was Hubble, nevertheless, who used the idea to develop a morphology that has been accepted.

Several authors, however, have credited both Hubble and Lundmark with independent development of the scheme.[43] Joint credit probably is allocated because Lundmark independently published a paper on classification[44] that was strikingly similar to Hubble's; indeed, Hubble believed that Lundmark had plagiarized his ideas, and complained about the matter to Slipher:

I see that Lundmark has published a "Preliminary Classification of Nebulae" which is practically identical with my own, except for the nomenclature. He calmly ignored my existence and claims it as his own exclusive idea. I am calling this to your official attention because I do not propose to let him borrow the results of hard labor in this casual manner.[45]

Hubble again mentioned the alleged plagiarism in a footnote in his paper of 1926,[46] in which he contended that since Lundmark was present at the 1925 I.A.U. meeting at Cambridge, where the classification was presented, Ludmark's scheme could not have been devised independently.

The validity of Hubble's claim has not yet been resolved, although two observations can be made. First, Lundmark's paper, which apparently was not based on the work of Jeans as was Hubble's, contained no appeal to an evolutionary sequence; and second, Lundmark had been working on classification at least as early as 1922, as is evidenced by a letter from him to Campbell:

Dr. Hale was very interested in my plans and he encouraged me to go ahead with certain questions about spiral nebulae. I will continue the work I started at Lick to get color equivalents for small spirals and secondary nuclei in large spirals. . . . Another piece of work will be the statistical investigation of known spirals in connection with the question of classifying the non-galactic nebulae.[47]

Since Lundmark had been considering a scheme, it is possible that he developed his ideas independently, and perhaps he followed the recommendations of the I.A.U. in presenting a purely descriptive model.

Lundmark commented on the accusation in a footnote to one of his papers in 1927:

In his paper, Extragalactic nebulae, Aph. J. 64:321, 1926, E. P. Hubble makes an attack on me which is written in such a tone that I hesitate to give any answer at all. Still, I may take the occasion to state a few facts.

I was present at the Cambridge meeting of the Astronomical Union.

I was not then a member of the Commission of Nebulae.

I did not have any access whatsoever to the memorandum or to other writings of E. P. Hubble, neither did I have access to the report of nebulae (which does not give details of Hubble's classification) until at the end of the meeting, neither did I recognize until I obtained a letter from Hubble at the end of 1926 that he had made another classification of nebulae than the one published in his paper, a general study of the Diffuse Galactic Nebulae, Mt. Wils. Contr. No. 241, 1922.

As much as I heard of the discussion in the committee of nebulae the only question was if the terms "galactic" and "extragalactic nebulae" should be accepted. From the discussion I got the impression that the intention of Hubble was to force through his nomenclature. One of the members told me outside the discussion that Hubble had suggested the subdivision "logarithmic spirals" but I did not understand that this suggestion was given in any memorandum to the Union. Now when reading Hubble's paper I am glad to note that he seems not to have carried out the unhappy idea introducing the term "logarithmic spirals." Slight changes in his classification might have been introduced since the Cambridge meeting.

Hubble's statement that my classification except for nomenclature is practically identical with the one submitted by him is *not correct*. Hubble classifies his subgroups according to excentricity [sic] or form of the spirals or degree of development while I use the degree of concentration towards the centre. As to the three main groups, elliptical, spirals and magellanic nebulae it may be of interest to note that the two first are slightly older than Hubble and myself. The term elliptical nebulae thus is used by Alexander in 1852 and the term spiral by Rosse in 1845. The importance of the magellanic group has been pointed out by myself (Observatory 47, 277, 1924) earlier than by Hubble. As to Hubble's way of acknowledging his predecessors I have no reason to enter upon this question here.[48]

165

The spiral galaxy NGC 4594 seen edge-on (Hale Observatories photograph)

Current Status

Aside from the premise of nebular evolution, Hubble's scheme for classifying non-galactic nebulae has become the accepted standard for most galaxies—most, because recent investigations indicate that some peculiar objects not yet well understood—such as exploding galaxies and Seyfert galaxies—do not seem to fit the Hubble system.

Although many revisions and other classification systems have been proposed since 1926,[49] Hubble's is still the most widely used.

Epilogue

for Section Three

The island-universe controversy, initiated in 1917 by the discovery of novae in spirals, was not finally resolved until 1935. The major stumbling block to the acceptance of the theory was van Maanen's measurements of internal motions in spirals, which indicated that the spirals could not be at the enormous distances required by the theory. It was only through the work of Hubble (in discovering Cepheids in spirals and in providing a basis for rejecting van Maanen's results) that the issue was ultimately resolved.

Although many astronomers contributed to our knowledge of galaxies during the 1920s, Hubble's achievements stand out beyond all others. The 1920s can truly be said to be Hubble's decade.

What we know today about galaxies we often take for granted. It seems, at first, that these things have been known for a long time. But now we have seen that the knowledge about galaxies—more important, the fact that other galaxies exist at all—is reasonably new. The whole of our knowledge today about galaxies and the structure of the universe has come about during the lifetime of the average man today. A man who is approximately 50 years old today was born before what we know as modern astronomy came into being.

167

References

Chapter 1

1 For biographical information about Slipher, see J. S. Hall, "V. M. Slipher's Trailblazing Career," *Sky and Telescope* 39 (February 1970): 84–86.
2 V. M. Slipher, "The Radial Velocity of the Andromeda Nebula," *Lowell Obs. Bull.* 58 (1914): 56–57.
3 A reprint of the paper presented at the AAS meeting is V. M. Slipher, "Spectrographic Observations of Nebulae," *Popular Astronomy 23* (1915): 21–24.
4 Private communication, E. Hertzsprung to V. M. Slipher, 14 March 1914 (Lowell Observatory Archives).
5 Slipher, "Spectrographic Observations."
6 "Dr. V. M. Slipher Tells of Inconceivable Distance of Dreyer Nebula No. 584," *The New York Times,* 19 January 1921, 6.

Chapter 2

1 V. M. Slipher, "The Detection of Nebular Rotations," *Lowell Obs. Bull.* 2 (1914): 66; "Spectrographic Observations of Nebulae," *Popular Astronomy 23* (1915): 21–24.
2 M. Wolf, *Astron. Gesell.* 49 (1914): 162.
3 H. D. Curtis, "Preliminary Note on Nebular Proper Motions," *Proc. Natl. Acad. Sci. USA 1* (1915): 10–12. Curtis later published a somewhat more detailed study: "Proper Motions of the Nebulae," *Publ. Astron. Soc. Pac. 27* (1915): 214–218.
4 Curtis, "Preliminary Note," 10.
5 *Ibid.,* 11.
6 *Ibid.,* 12.
7 A. van Maanen, "Preliminary Evidence of Internal Motion in the Spiral Nebula Messier 101," *Astrophys. J. 44* (1916): 210–228; "Preliminary Evidence of Internal Motion in the Spiral Nebula Messier 101," *Proc. Natl. Acad. Sci. USA 2* (1916): 386–390.
8 The results were summarized in A. van Maanen, "Internal Motions in Four Spiral Nebulae," *Publ. Astron. Soc. Pac. 33* (1921): 200–202. The measures of each spiral were published separately: "Preliminary Evidence of Internal Motion in Spiral Nebula Messier 101"; "Investigations on Proper Motion, Fourth Paper: Internal Motions in the Spiral Nebula Messier 51," *Astrophys. J. 54* (1921): 237–245; "Investigations on Proper Motions, Fifth Paper: Internal Motions in the Spiral Nebula Messier 81," *Astrophys. J. 54* (1921): 347–356; "Internal Motion in the Spiral Nebula Messier 33," *Proc. Natl. Acad. Sci. USA 7* (1921): 1–5.

9 Private communication, J. C. Duncan to V. M. Slipher, 14 July 1916 (Lowell Observatory Archives).
10 Private communication, A. van Maanen to H. Shapley, 5 June 1921 (Harvard University Archives).
11 Private communication, A. van Maanen to G. E. Hale, 11 July 1917 (Hale Collection).
12 It is surprising that van Maanen was unable to find motions in M31, which we now know to be the nearest spiral, whereas he was able to find motions in others, which are immensely more distant.
13 Private communication, A. van Maanen to G. E. Hale, 17 December 1917 (Hale Collection).
14 A. van Maanen,. "Messier 33," (1921): 1.
15 *Ibid.*
16 *Cf.* references 7 and 8. A. van Maanen, "Investigations on Proper Motion, Seventh Paper: Internal Motions in the Spiral Nebula NGC 2403," *Astrophys. J. 56* (1922): 200–207; "Eighth Paper: Internal Motions in the Spiral Nebula M94=NGC 4736," *Astrophys. J. 56* (1922): 208–216; "Ninth Paper: Internal Motions in the Spiral Nebula Messier 63, NGC 5055," *Astrophys. J. 57* (1923): 49–56; "Tenth Paper: Internal Motions in the Spiral Nebula Messier 33, NGC 598," *Astrophys. J. 57* (1923): 264–278.
17 A. van Maanen, Tenth Paper.
18 H. Shapley, "On the Existence of External Galaxies," *Publ. Astron. Soc. Pac. 31* (1919): 261–268.
19 H. Shapley, *Through Rugged Ways to the Stars* (New York: Scribners, 1969), 79.
20 Private communication, H. Shapley to H. N. Russell, 31 March 1920 (Harvard University Archives).
21 Private communication, H. Shapley to H. N. Russell, 3 September 1917 (Harvard University Archives). Note that Shapley mentioned results for M33 that van Maanen had not yet published.
22 Private communication, H. N. Russell to H. Shapley, 8 November 1917 (Princeton University Archives).
23 Private communication, H. Shapley to H. D. Curtis, 24 October 1920 (Harvard University Archives).
24 *Cf.* reference 10.
25 Private communication, H. Shapley to A. van Maanen, 8 June 1921 (Harvard University Archives).
26 Private communication, A. van Maanen to H. N. Russell, 7 September 1920 (Princeton University Archives).

27 Private communication, H. N. Russell to A. van Maanen, 5 October 1920 (Princeton University Archives).

Chapter 3

1 Private communication, J. Stebbins to H. D. Curtis, 22 January 1924 (Allegheny Observatory Archives).
2 Private communication, H. D. Curtis to J. Stebbins, 27 January 1924 (Allegheny Observatory Archives).
3 Ibid.
4 Private communication, V. M. Slipher to H. D. Curtis, 10 June 1924 (Lowell Observatory Archives).
5 Private communication, H. D. Curtis to W. H. Burtt, 25 January 1922 (Allegheny Observatory Archives).
6 Private communication, H. D. Curtis to P. O'Dea, 5 July 1924 (Allegheny Observatory Archives).
7 E. P. Hubble, "The Direction of Rotation in Spiral Nebulae," Astrophys. J. 97 (1943): 112–118. See also K. J. Gordon, "History of Our Understanding of a Spiral Galaxy: Messier 33," Q. J. R. Astron. Soc. 10 (1969): 302.
8 The details of that announcement and insight into van Maanen's influence on Hubble are given in: R. Berendzen and M. Hoskin, "Hubble's Announcement of Cepheids in Spiral Nebulae," Leaf. Astron. Soc. Pac. (June 1971); M. A. Hoskin, "Edwin Hubble and the Existence of External Galaxies," paper delivered at the XIIᵉ Congrès International d'Histoire des Sciences, Paris, 1968.
9 Private communication, H. D. Curtis to R. G. Aitken, 2 January 1925 (Lick Observatory Archives).
10 Private communication, A. van Maanen to H. Shapley, 5 June 1921 (Harvard University Archives).
11 Private communication, J. C. Duncan to V. M. Slipher, 14 July 1916 (Lowell Observatory Archives).
12 Jeans considered spirals to be merely one stage in an evolving nebula. He believed that an initially spherical mass of gas in rotation would flatten as it contracted and eventually become unstable, ejecting material in filaments from its edges, thereby forming spiral arms. See, for example, J. Jeans, Problems of Cosmogony and Stellar Dynamics (Cambridge, Eng.: Cambridge University Press, 1919).
13 Private communication, A. van Maanen to H. Shapley, 22 June 1921 (Harvard University Archives).
14 Private communiction, A. van Maanen to H. Shapley, 17 August 1921 (Harvard University Archives).
15 A. van Maanen, "Internal Motions in Four Spiral Nebulae," Publ. Astron. Soc. Pac. 33 (1921): 200–202.
16 Ibid., 202.
17 J. Jeans, "Internal Motion in Spiral Nebulae," Observatory 40 (1917): 60–61.

18 "Proceedings at the Meeting of the Royal Astronomical Society," Observatory 44 (1921): 353.
19 Ibid., 355.
20 A. S. Eddington, "The Motions of Spiral Nebulae," Mon. Not. R. Astron. Soc. 77 (1917): 377.
21 J. H. Reynolds, "Nebulae," Mon. Not. R. Astron. Soc. 84 (1924): 285.
22 J. H. Jeans, "Internal Motions in Spiral Nebulae," Mon. Not. R. Astron. Soc. 84 (1923): 60–76.
23 Ibid., 60.
24 Ibid., 67.
25 Ibid., 72. It should be noted that the temper of the times was highly suitable for such an apparently radical proposal, for it followed two of the most revolutionary decades in science. Unconventional proposals in physics were becoming commonplace.
26 Ibid., 76.
27 J. H. Jeans, "Note on the Distances and Structure of the Spiral Nebulae," Mon. Not. R. Astron. Soc. 85 (1925): 531.
28 These notes were also attached to a letter from J. H. Jeans to H. N. Russell, 23 October 1924 (Princeton University Archives). This, incidentally, provides proof of the widespread dissemination of knowledge of Hubble's discovery before the public announcement of the work at AAS Washington meeting in December 1924. Further details can be found in the papers iñ reference 8 above. Although Jeans had to give up dependence on van Maanen as an observational champion of his own work, he soon found another in Hubble. Hubble's work on the classification of galaxies was adopted by Jeans as the observational evidence for his own work.
29 In the early 1920s the most prominent astrophysicists in the world were A. S. Eddington and J. H. Jeans, both Englishmen. Consequently, England was the principal center of influence in theoretical astronomy.
30 E. W. Brown, "Gravitational Forces in Spiral Nebulae," Astrophys. J. 61 (1925): 97–113.
31 E. W. Brown, "Gravitational Motion in a Spiral Nebula," Observatory 51 (1928): 278.
32 C. O. Lampland, "On the Proper Motion of the Virgo Nebula, NGC 4594," Popular Astronomy 22 (1914): 631–632.
33 C. O. Lampland, "Preliminary Measures of the Spiral Nebulae NGC 5194 (M51) and NGC 4254 (M99) for Proper Motion and Rotation," Popular Astronomy 24 (1916): 667–668.
34 C. O. Lampland, "On Changes Observed in the Nucleus of the Spiral Nebula NGC 4254 (Messier 99)," Publ. Astron. Soc. Pac. 33 (1921): 167–168.
35 Private communication, H. Shapley to A. van Maanen, 17 May 1921 (Harvard University Archives). Shapley must have had private correspondence with Lampland since Lampland's paper (reference 34 above) was dated 18 May.

36 By 1921, Curtis and Lundmark were almost alone in openly opposing van Maanen. Not only did van Maanen have Shapley, Russell, and Jeans on his side, but also much of the rest of the astronomical community. After returning from a meeting (probably AAS), Shapley wrote to van Maanen (8 September 1921, Harvard University Archives):

> I think that your nebular motions are taken seriously now, and nobody but Very dared raise his head after I explained how dead the island universes are if your measures are accepted. And he later came around.

37 Private communication. A. van Maanen to H. Shapley, 23 May 1921 (Harvard University Archives).

38 Private communication, E. P. Hubble to V. M. Slipher, 24 July 1923 (Lowell Observatory Archives).

39 S. Kostinsky, "Probable Motions in the Spiral Nebula Meissier 51 (Canes Venatici) Found with the Stereo-comparator, Preliminary Communication," *Mon. Not. R. Astron. Soc.* 77 (1917): 223–224.

40 W. J. A. Schouten, "Probable Motions in the Spiral Nebula Messier 51 (Canes Venatici)," *Observatory* 42 (1919): 441–444.

41 A. van Maanen, "Investigations on Proper Motion, Tenth Paper: Internal Motions in the Spiral Nebula Messier 33, NGC 598," *Astrophys. J.* 57 (1923): 264–278.

42 H. Shapley and H. D. Curtis, "The Scale of the Universe," *Bull. Natl. Res. Coun.* 2 (1921): 214.

43 H. D. Curtis, "The Nebulae," *The Adolpho Stahl Lectures in Astronomy* (San Francisco: Stanford University Press, 1919): 95–107.

44 Private communication, H. D. Curtis to W. W. Campbell, 11 July 1922 (Lick Observatory Archives).

45 *Ibid.*

46 Private communication, A. van Maanen to W. W. Campbell, 9 June 1922 (Lick Observatory Archives).

47 Private communication, W. W. Campbell to H. D. Curtis, 12 June 1922 (Lick Observatory Archives).

48 Private communication, K. Lundmark to W. W. Campbell, 13 June 1922 (Lick Observatory Archives).

49 Private communication, H. D. Curtis to A. van Maanen, 28 January 1925 (Allegheny Observatory Archives).

50 Letters of H. D. Curtis to R. Merrill, 8 January 1925; to Shock, 24 March 1925; to S. Kuftinec, 12 April 1925 (Allegheny Observatory Archives). Also, he wrote to W. M. Smart in England on 13 May 1924 pointing out the apparent contradiction in direction of rotation and asking the " 'mathematical sharks' on the other side of the water" to consider the difficulty (Allegheny Observatory Archives). Smart was an enthusiastic supporter of van Maanen's results, as he had recently stated in a lengthy paper: "The Motions of Spiral Nebulae," *Mon. Not. R. Astron. Soc.* 84 (1924): 333–353.

51 H. D. Curtis, "The Nebulae," *Handb. Astrophys.* 5 (Berlin 1933): 774–936.

52 Cf. ref 37.

53 K. Lundmark, "The Spiral Nebula Messier M33," *Publ. Astron. Soc. Pac.* 33 (1921): 324–327.

54 K. Lundmark, "The Proper Motions of Spiral Nebula," *Popular Astronomy* 30 (1922): 623.

55 K. Lundmark, "On the Motion of Spirals," *Publ. Astron. Soc. Pac.* 34 (1922): 108–115.

56 *Ibid.*, 109.

57 Private communication, K. Lundmark to R. G. Aitken, 31 May 1924 (Lick Observatory Archives).

58 K. Lundmark, "Internal Motions of Messier 33," *Astrophys. J.* 63 (1926): 67.

59 H. Shapley, *Through Rugged Ways to the Stars* (New York: Scribners, 1969): 80–81.

60 J. H. Jeans, *Astronomy and Cosmogony* (Cambridge, Eng.: Cambridge University Press, 1928): 351.

61 K. Lundmark, "Studies of Anagalactic Nebulae: First Paper," *Ups. Astron. Obs. Med.* 30 (1927): 48–49.

62 A. van Maanen, Investigations on Proper Motion, Eighth Paper: Internal Motions in the Spiral Nebula M 94=NGC 4736," *Astrophys. J.* 56 (1922): 208–216. The attack was probably in response to Lundmark's criticisms in his 1922 paper (see reference 55).

63 Lundmark's original statements about spirals are included in his exhaustive dissertation: "The Relations of the Globular Clusters and Spiral Nebulae to the Stellar System," *Kungl. Sven. Veten. Handl.* 60 (1920): 1–79. Van Maanen criticized Lundmark's assumption that spirals are comparable to the Milky Way and the Magellanic clouds; he asserted that this was tantamount to begging the question.

64 A. van Maanen, "M94," (1922): 216.

65 It is possible that Lundmark could have discussed qualitative agreement with van Maanen before returning to Uppsala.

66 A. van Maanen, "Messier 33," (1923): 278.

Chapter 4

1 For more information concerning Leavitt's discovery, see H. Shapley, *Star Clusters* (New York: McGraw-Hill, 1930): 125–128.

2 E. Hubble, "Cepheids in Spiral Nebulae," *Observatory* 48 (1925): 140.

3 Private communication, E. Hubble to H. Shapley, 19 February 1924 (Henry E. Huntington Library Archives).

4 Private communication, H. Shapley to E. Hubble, 27 February, 1924 (Harvard University Archives). This letter was brought to the author's attention by O. Gingerich.

5 Private communication, H. Shapley to G. E. Hale, 6 March 1924 (Hale Collection).

6 Private communication, H. N. Russell to W. Davis, 12 December 1924 (Princeton University Archives).

7 Private communication, H. N. Russell to E. Hubble, 12 December 1924 (Henry E. Huntington Library Archives).

8 R. Berendzen and M. Hoskin, "Hubble's Announcement of Cepheids in Spiral Nebulae," *Leaf. Astron. Soc. Pac.* (June 1971): 5–7.

9 A. Sandage, *The Hubble Atlas of Galaxies* (Washington: Carnegie Institute, 1961): 4–5.

10 Private communication, J. Stebbins to the Committee on the Second A.A.A.S. Prize, 1 January 1925 (American Institute of Physics Archives).

11 Private communication, E. Hubble to H. N. Russell, 19 February 1925 (Princeton University Archives).

12 "Dr. E. Hubble Confirms View that Spiral Nebuale Are Stellar Systems," *The New York Times*, 23 November 1924, 6.

13 Private communication, H. D. Curtis to S. Kuftinec, 12 April 1925 (Allegheny Observatory Archives).

Chapter 5

1 For details, see J. D. Fernie, "The Period-Luminosity Relation: A Historical Review," *Publ. Astron. Soc. Pac. 81* (1969): 707–731.

2 A. van Maanen, "Investigations on Proper Motion, Tenth Paper: Internal Motions in the Spiral Nebula Messier 33, NGC 598," *Astrophys. J. 57* (1923): 274.

3 *Ibid.,* 275.

4 A. van Maanen, "Investigations on Proper Motion, Eleventh Paper: The Proper Motion of Messier 13 and Its Internal Motion," *Astrophys. J. 61* (1925): 130–136.

5 A. van Maanen, "Investigations on Proper Motion, Twelfth Paper: The Proper Motions and Internal Motions of Messier 2, 13, and 56," *Astrophys. J. 66* (1927): 89.

6 A summary of the meeting is given in *Mon. Not. R. Astron. Soc. 85* (1925): 897–903.

7 The discovery of the velocity-distance relation by Hubble in 1929, which fitted so well with the theoretical considerations of de Sitter and others, obviously required spirals to be distant.

8 A. van Maanen, "Investigations on Proper Motion, Fifteenth Paper: The Proper Motion of the Spiral Nebula NGC 4051," *Mt. Wilson Contributions 407* (1930): 1–6.

9 Private communication, P. W. Merrill to H. D. Curtis, 2 November 1924 (Allegheny Observatory Archives).

It is unfortunate for the historian that both Hubble and van Maanen worked at Mount Wilson, for no correspondence between them seems to exist. But according to Shapley, they never liked one another; thus they might not have communicated under any condition. Shapley later wrote in *Through Rugged Ways to the Stars* (New York: Scribners, 1969), 80–81:

> Adrian [sic] van Maanen came to Mount Wilson within two or three years after I did. He at once became a friend of Mr. Hale. Van Maanen was aggressive and he was sociable. He could go to a dinner and soon have the whole table laughing. He was a social success. People liked him—until he became sort of a playboy.

Van Maanen's job at Mount Wilson was measuring galaxies, and stellar parallaxes, and the sun's magnetic field by way of the Zeeman effect. He worked for Hale and was soon getting the rotation of galaxies. It looks now as though he got the answers he wanted, or that seemed to be best. I don't know that he ever corrected himself, but others have corrected him. He was a charming person, a bachelor; he and I were pals of a sort—I don't know why, because I wasn't "society" and he was. I suppose we got together because he was rather an alert-minded person and I liked his nonsense. Hubble disliked van Maanen from the time he himself arrived on Mount Wilson; he scorned him. Hubble just didn't like people. He didn't associate with them, didn't care to work with them. I remember once somebody referred a paper of mine to Hubble for him to pass judgment on. It was a good paper; it was correct; I mean I knew what I was talking about at that time. It was written for some journal like *Scientific American*. Hubble just wrote across it, "Of no consequence." The editors, who told me about it thought it was the funniest thing, because the words, "Shapley—of no consequence" got set in type.

Hubble and I did not visit very much. He was a Rhodes scholar, and he didn't live it down. He spoke with a thick Oxford accent. He was born in Missouri not far from where I was born and probably knew the Missourian tongue. But he spoke "Oxford." He would use such phrases as "to come a cropper." The ladies he associated with enjoyed that Oxford touch very much. "Bah Jove!" he would say, and other such expressions. He was quite picturesque.

10 Private communication, R. G. Aitken to H. D. Curtis, 26 January 1925 (Allegheny Observatory Archives).

11 Private communication, H. N. Russell to E. P. Hubble, 12 December 1924 (Henry E. Huntington Library).

12 Private communication, H. N. Russell to W. Davis, 12 December 1924 (Princeton University Archives).

13 H. N. Russell, "The Nebulae," Lectures at the University of Toronto, February 1924 (unpublished transcript in the Princeton University Archives).

14 H. D. Curtis, "The Nebulae," *Handb. Astrophys. 5* (Berlin 1933): 851.

15 A. van Maanen, "Spiral Nebula NGC 4051," 6.

16 E. P. Hubble, "NGC 6822, A Remote Stellar System," *Astrophys. J. 62* (1925): 409–433; "A Spiral Nebula as a Stellar System: Messier 33," *Astrophys. J. 63* (1926): 236–274; "A Spiral Nebula as a Stellar System, Messier 31," *Astrophys. J. 69* (1929): 103–157.

17 E. P. Hubble, "Angular Rotations of Spiral Nebulae," *Astrophys. J. 81* (1935): 334–335.

18 A. van Maanen, "Internal Motions in Spiral Nebulae," *Astrophys. J. 81* (1935): 336–337.

19 *Ibid.,* 337.

20 J. D. Fernie, "The Historical Quest for the Nature of the Spiral Nebulae," *Publ. Astron. Soc. Pac. 82* (1970): 1219.

21 W. Baade, *Evolution of Stars and Galaxies* (Cambridge: Harvard University Press, 1963): 28–29.

22 N. S. Hetherington, "Adriaan van Maanen and Internal Motions in Spiral Nebulae: A Historical Review," *Q. J. R. Astron. Soc. 13* (1972): 25–39.

23 Shapley, *Through Rugged Ways to the Stars*, 56–57.

24 For example, see Hetherington, "Adriaan van Maanen."

25 R. Hart, "Adriaan van Maanen's Influence on the Island Universe Theory" (Dissertation, Boston University, 1973).

26 A. van Maanen, "Internal Motions in Spiral Nebulae," 337.

27 See, for example, B. Lindblad, "On the Spiral Orbits in the Equitorial Plane of a Spheroidal Disk with Applications to Some Typical Spiral Nebulae," *Ups. Astrom. Obs. Med. 31* (1927): 1–18.
Since van Maanen never directly mentioned Lindblad, it is impossible to assess the effect of this theoretical work on him; however, van Maanen probably was aware of it. Also, Lundmark mentioned Lindblad's theory several times in his 1927 paper, which may make his failure to press home the attack against van Maanen more understandable, as the theory supported van Maanen's direction for the rotations.

28 For a detailed study of absorption, see D. Seeley and R. Berendzen, "The Development of Research in Interstellar Absorption, c. 1900–1930," *J. Hist. Astron. 3* (1971): 52–64, 75–86.

29 Private communication, E. P. Hubble to V. M. Slipher, 11 June 1941 (Oort Papers, Leiden Observatory Archives).

30 E. P. Hubble, The Direction of Rotation in Spiral Nebulae," *Astrophys. J. 97* (1943): 112–118.

31 It is interesting to speculate why the discovery was not made sooner. By 1913, Cepheids had been identified in the Magellanic Clouds and, by 1918, the period–luminosity relation (with which distances to Cepheids could be determined) was available. Thus, after 1918, someone might reasonably have attempted to find a Cepheid in a spiral. Certainly, capable interested observers were present, as Fernie noted ("Historical Quest," 1226):
Shapley himself, so adept at finding variable stars in globular clusters, had been in an excellent position for investigating their possible existence in spiral nebulae, but apparently so convinced was he that the nebulae were altogether something else, that he never did so.
(Shapley maintained [*Through Rugged Ways to the Stars*, 57–58], however, that he did not do so because it was not his assigned task at Mount Wilson.)

Even when variable stars were found in M33 (before Hubble's discovery), apparently no one suggested that they might be Cepheids. (J. C. Duncan, "Three Variable Stars and a Suspected Nova in the Spiral Nebula M33 Trianguli," *Publ. Astron. Soc. Pac. 34* (1922): 290–291; M. Wolf, "Zwei Neue Veranderliche, Trianguli, im Spiral Nebel M 33," *Astron. Nachr. 217* (1923): 475.)
And when Hubble made the definitive discovery in 1923, he at first thought he had found a nova instead of a variable, as is clear from his notebook entries and plate markings. (See N. U. Mayall, "Edwin Hubble: Observational Cosmologist," *Sky and Telescope 13* (1954): 78.) Thus, the discovery was accidental, not the result of a search.
It is interesting to note that apparently no one proposed searching for Cepheids as a diagnostic of the island-universe theory and, indirectly, the period-luminosity relation. There are several possible reasons for this: Shapley's P-L relation was not accepted universally; in the entire world, only a handful of observers and telescopes were involved in studies of spirals; and hindsight may make the problem appear more obvious today than it was then. But it is possible that van Maanen's results, which seemed so lethal to the theory, may have dissuaded some astronomers from considering such an idea.

Chapter 6

1 The details of William Herschel's scheme are described by H. D. Curtis in "Nebulae," *Handb. Astrophys. 5* (Berlin, 1933): 919.

2 M. Wolf, "Die Klassifizierung der Kleinen Nebelflecken," *Publ. Astrophys. Inst. Koingstuhl-Heidelberg 3* (1909): 109–112.

3 E. Hubble, "Photographic Investigations of Faint Nebulae," *Publ. Yerkes Obs. 4* (1920): 69.

4 E. Hubble, "A General Study of Diffuse Galactic Nebulae," *Astrophys. J. 56* (1922): 162.

5 *Ibid.*, 166.

6 *Ibid.*, 168.

7 More information about the meeting can be found in *Observatory 45* (1922): 176–190; and in *Publ. Astron. Soc. Pac. 34* (1922): 275–285.

8 "The American Section of the International Astronomical Union—Report of the Committee on Nebulae," submitted March 1922 (Lowell Observatory Archives). The members of the American Section were E. E. Barnard, E. Hubble, C. O. Lampland, V. M. Slipher, and V. H. Wright.

9 The report is published in *International Astronomical Union Transactions 1* (1922): 91–94.

10 Bigourdan had published this scheme previously in *C.R. Acad. Sci 158* (1914): 1949–1957.

11 Private communication, K. Lundmark to W. W. Campbell, 13 June 1922 (Lick Observatory Archives).

12 Background material on Hubble is given by N. U. Mayall, "Edwin Hubble Biographical Memoir for the National Academy of Sciences," December 1966; also, N. U. Mayall, "Edwin Hubble, Observational Cosmologist," *Sky and Telescope* (January 1954): 78–85.

13 Hubble, "Diffuse Galactic Nebulae," 167.

14 "American Section Report," (1922) 5.

15 J. Jeans, *Problems of Cosmology and Stellar Dynamics* (Cambridge, Eng.: Cambridge University Press, 1919).

16 Private communication, E. Hubble to V. M. Slipher, 4 April 1923 (Lowell Observatory Archives).

17 *Ibid.*

18 H. Jeffreys, "On Jeans' Theory of the Origin of Spiral Nebulae," *Mon. Not. R. Astron. Soc. 83* (1923): 449–453. The purpose of this paper was to criticize a mathematical point of Jeans' theory. At the end of the paper he admits, however, that, having seen Jeans' reply, the criticism was not valid and the theory as stated was correct.

19 Private communication, E. Hubble to V. M. Slipher, 24 July 1923 (Lowell Observatory Archives).

20 E. Hubble, "The Classification of Nebulae" (Lowell Observatory Archives). This is the manuscript that Hubble sent to V. M. Slipher in 1923. It was later distributed to the members of the I.A.U. Commission on Nebulae.

21 *Ibid.*

22 *Ibid.*

23 The eccentricity of elliptical nebulae was to be found by comparing the photographic image with a series of ellipses of eccentricities 0.0, 0.2, 0.4, 0.6, 0.8. The decimal point was dropped and n was expressed as an integer between 0 and 8.

24 The logarithmic spiral is now called the "normal" spiral.

25 The barred spiral was the ϕ shaped spiral previously mentioned by Hubble. This shape was first mentioned by H. D. Curtis in the *Publ. Lick Obs. 13* (1918): 12.

26 Hubble, "Classification."

27 *Ibid.*

28 Hubble to Slipher, 24 July 1923.

29 Private communication, E. Hubble to V. M. Slipher, 9 February 1924 (Lowell Observatory Archives).

30 Holetschek, *Annuales der Wierner Sternwurk 20* (1907), a catalog of total luminosities for 417 extragalactic nebulae.

31 Hubble explained this relation more fully in his 1926 paper, "Extra–Galactic Nebulae," *Astrophys. J. 64* (1926): 321–369.
> The distribution of magnitudes appears to be uniform throughout the sequence. For each type or stage in the sequence, the total magnitudes are related to the logarithms of the maximum diameters by the formula
> $$M_T = C - 5 \log d,$$
> where C varies progressively from type to type, indicating a variation in diameter for

a given magnitude or vice versa. By applying corrections to C, the nebulae can be reduced to a standard type and then a single formula expresses the relaxation for all nebulae from the Magellanic clouds to the faintest that can be classified. . . . The coefficient of log d corresponds with the inverse square law, which suggests that the nebulae are all of the same order of absolute luminosity and that apparent magnitudes are measures of distance.

32 Private communication, E. Hubble to V. M. Slipher, 22 June 1926 (Lowell Observatory Archives).

33 E. Hubble, "Extra-Galactic Nebulae," 321–369

34 The changes were trivial: n was allowed to take the integral values between 0 and 7, instead of between 0 and 8 as in the 1923 manuscript ("Classification"), and a few of the examples were changed.

35 The "tuning fork" diagram, which was a slightly revised version of the 1926 ordering, was used by Hubble in *The Realm of the Nebulae* (New Haven: Yale University Press, 1936), 45. Actually, the "tuning fork" closely resembles the Y diagram that had been used by Jeans in *Astronomy and Cosmology* (Cambridge, Eng.: Cambridge University Press, 1928), 324, in which he attempted to show that this theory of evolution for spirals was related to Hubble's observations.

36 Hubble, "Extra-Galactic Nebulae," 346.

37 *Ibid.*, 324.

38 It was decided at the 1922 Rome meeting that the Commission should also consider star clusters.

39 *International Astronomical Union Transactions 2* (1925): 206.

40 Private communication, 13 April 1970. From its origin until recently, Dr. Mayall was Director of the Kitt Peak National Observatory.

41 J. Jeans, "An Evolving Universe," *Carnegie Institution of Washington, News Service Bulletin, Staff Edition 23* (1931): 157.

42 H. N. Russell, "The Nebulae," lecture given at the University of Toronto, February 1924: 260 (Russell Collection, Princeton University Archives).

43 See, for example, L. Motz and A. Duveen, *Essentials of Astronomy* (Belmont, Calif., Wadsworth Publ. Co., 1967): 569; or W. Baade, *Evolution of Stars and Galaxies* (Cambridge: Harvard University Press, 1963): 12.

44 K. Lundmark, "A Preliminary Classification of Nebulae," *Ark. Math. Astron. Fys. 19* B (1926). This paper was published several months before Hubble's article of 1926 appeared in print.

45 Hubble to Slipher, 22 June 1926.

46 Hubble, "Extra-Galactic Nebulae," 323.

47 Private communication, K. Lundmark to W. W. Campbell, 28 May 1922 (Lick Observatory Archives).

48 K. Lundmark, "Studies of Anagalactic Nebulae, First Paper," *Med. Astron. Obs. Ups. 30* (1927), 24.

49 Some important revisions and classifications include H. Shapley, "On the Classification of Extra-Galactic Nebulae," *Harvard Bull. 849* (1927). It is interesting to note that while in the 1927 paper Shapley severely criticized Hubble's scheme and proposed one of his own, in "Second Note on the Relative Number of Spiral and Elliptical Nebulae," *Harvard Bull. 876* (1930), he used Hubble's morphology in his discussion; W. W. Morgan, "A Preliminary Classification of the Forms of Galaxies According to Their Stellar Population," *Publ. Astron. Soc. Pac. 70* (1958): 394; G. de Vaucouleurs, "Classification and Morphology of External Galaxies," *Handb. Phys. 53* (1959): 275–372; S. Van den Burgh, "A Preliminary Luminosity Classification of Late-Type Galaxies," *Astrophys. J. 131* (1960): 215; A. R. Sandage, *The Hubble Atlas of Galaxies* (Washington: Carnegie Institute, 1961).

FOUR

The Birth
of Modern
Cosmology

Contents

of Section Four

Albert Einstein.

Prologue
for Section Four

Cosmology, in the past, has often been considered a branch of metaphysics; philosophers were the persons who delved into the structure and order of the universe. Only recently have astronomers begun to test their models observationally. That does not mean that certain cosmological problems were not recognized before the advent of large telescopes, which enabled astronomers to penetrate deep into space and far back in time, nor does it mean that they were unsuccessful in proposing solutions to their cosmological dilemmas. As the first chapter of this Section demonstrates, certain difficulties associated with an infinite universe were considered centuries ago.

The theory of gravitation proposed by Newton in the seventeenth century was very successful in dealing with planetary motions. Difficulties arose, however, when it was applied to the entire universe—gravity required the universe to be unstable. A resolution of the unsatisfactory elements resulted from the relativistic treatment of gravitation by Albert Einstein around 1917.

177

The introduction of relativistic physics by Einstein, and its subsequent acceptance by many scientists, led to new approaches to the structure of space and new cosmologies. Both Einstein and Willem de Sitter found solutions to Einstein's field equations that expressed the form of space in mathematical terms; both solutions were based on the assumption that our universe was static. These first relativistic cosmologies offered a resolution of the problems associated with Newtonian mechanics and revived astronomical interest in the study of the cosmos.

The relativistic cosmologies developed during the period 1905-1930 were based solely on philosophical and theoretical considerations because relevant observations were lacking. Certain gaps in astronomical knowledge were filled during the 1920s, however, with far-reaching consequences on the early cosmologies. Once the basic structural units of our universe had been determined and the problem of scale solved, astronomers were able to address themselves to new questions.

When it became known from observation that spiral nebulae are, in fact, star systems similar to the Milky Way and at large distances from it, the question of the relation among such objects in space became important. The large radial velocities of the spirals discovered by Vesto M. Slipher were the outstanding observations that required explanation.

Although distances to a few spirals had been determined by the late 1920s, determining distances was still a major difficulty in attempting to recognize order in the large-scale structure of the universe. Several attempts had been made to relate radial velocities of spirals to their distances, but all were unsuccessful until Hubble's linear relation between spectral redshift and distance was published in 1929.

Theoretical attempts at constructing such a relationship—although formulated as early as 1917—were also unsuccessful until the observations became more reliable. Finally, in the early 1930s, observations and theories describing an expanding universe joined to formulate a view of our universe that is still acceptable in principle today. The details of such a universe, however, are almost constantly changing as instrumentation, techniques, and theories become more refined, and new discoveries are made. The accepted size of the universe has been steadily increasing since the 1930s, and probably will increase yet further even as this is being written.

A 1916 photograph of a group of spiral nebulae taken with Mount Wilson's 60-inch telescope. (Hale Observatories photograph)

CHAPTER 1

Cosmological Problems

Olbers' Paradox

Have you noticed that the sky is dark at night? Of course we can perceive the light of stars, planets, the moon, and often the glow of city lights reflected from particulate matter in our atmosphere, but one cannot deny that nighttime is darker than daytime. Why, *naturally* the daytime is bright, you might reply; the Sun illuminates the Earth because it is near, but the stars are so distant that their light is negligible. In general, people would agree that the difference between night and day is not strange at all and hardly worth serious discussion.

In spite of this general consensus, the darkness of the nighttime sky and similar problems have perplexed philosophers for centuries. Before examining the dilemmas encountered and the attempts at their resolution, let us consider the geometric aspects of the problem. Let us divide the universe into concentric shells, like the layers of an onion, each having a thickness t; the volume of each shell will be approximately the surface area of one of the boundary concentric spheres times t. From elementary geometry, we know that the surface of a sphere increases in proportion to the square of its radius (exactly, the surface area A of a sphere of radius R is expressed as $A = 4\pi R^2$). Consequently, the volume of each shell, and hence the number of stars in each shell, is proportional to the square of the radius. For each star in a given shell, the brightness is proportional to $1/R^2$; but since the number of stars is proportional to R^2, the total amount of light reaching us from all the stars in any shell is independent of the distance of that shell—each shell contributes an equal amount to the luminosity of the sky! The fact that the stars are not perfectly uniform in their distribution in space does not alter the argument significantly; the fluctuations average out if we consider many shells.

The paradoxical consequence of these arguments comes when the light we see from the various shells is added together. As R increases, incorporating more and more shells, the light received should increase *ad infinitum* until the entire sky should be ablaze, as is the surface of the Sun, irrespective of the time of day. Of course, that does not happen, but the arguments are logical. Even the fact that the stars seen are restricted to the finite volume of one galaxy does not resolve the dilemma; the same arguments can be applied to the galaxies as well as to the stars. The darkness of the nighttime sky is definitely a significant, and perplexing, cosmological problem.

Before the seventeenth century, most philosophers ignored this problem because they possessed a deficient concept of the manner in which light affects the human senses. They thought, correctly, that the light of an individual star will be diminished to such an extent that beyond a certain distance the eye can not receive an individual optical impression. Therefore they concluded, incorrectly, that any stars that might exist, beyond the few thousand that can be seen with the naked eye, could have no effect on the light received on the Earth. Since any contribution that by itself was insignificant could be equated with zero, the sum of many undetectable rays of light was still equal to zero. From such a view, the paradox could not have been conceived by the early savants. Not until after Galileo looked at the Milky Way through his telescope was this mistake recognized.

One of the many celestial objects that impressed Galileo when he turned his telescope towards the sky was the Milky Way. For centuries people had debated the nature of the nebulous band in the sky and many had guessed that the light was stellar, but, upon inspection with a telescope, Galileo knew for sure that there were innumerable faint stars clustered together. Furthermore, the telescope revealed to Galileo that other regions classified as "nebulous" were, in fact, clusters of stars. The implications of these observations did not escape him; he wrote: "Although each star separately escapes our sight on account of its smallness or the increased distance from us, the mingling of their rays gives rise to that gleam which was formerly believed to be some denser part of the aether that was capable of reflecting rays from stars or from the sun."[1] This insight into the optical effects of many exceedingly faint stars provided a necessary step towards the recognition of the paradox.

Nevertheless, Galileo was unable to free himself completely from the previous philosophical background in regard to the detection of light. Although he realized that the "mingling of their rays" allowed the imperceptible stars of the Milky Way to appear as a nebulous band, he did not apply the same idea to more distant stars. Beyond a certain distance, according to Galileo, stars had no effect upon man's vision. Furthermore, he expressed the notion that stars did not exist beyond a certain distance. Hence, no paradox existed in his cosmology.

Early recognition of the paradox came around 1720. Edward Halley, who became famous for his comet predictions, described the comments of persons (whom he did not identify) who had stated that a finite universe was necessary to avoid the existence of a sky as bright as the Sun in every direction. Here is a definite expression of the dilemma stated previously—an infinite universe implies a bright sky at all times of day. Although the unnamed person provided a simple resolution of the problem, in the form of a finite universe, Halley was not receptive. He himself could not relinquish his firm belief in an infinite universe; yet he could not express a coherent counterargument.

Jean-Phillipe Loys de Chéseaux, a young mathematical genius, reiterated the problem about 1740 and also proposed solutions to it. Like the unnamed person

181

W.H.M. Olbers. (Sky Publishing Co. photograph)

mentioned by Halley, Chéseaux also considered the possibility of a finite universe. He was not satisfied with that explanation, however, and as an alternative solution proposed the diminution of the light intensity because of absorption along its path. This possibility would be reasonable if the interstellar space was filled with some substance that lacked perfect transparency. Chéseaux postulated that even a substance 3.3×10^{17} times more transparent than water would be sufficient to reduce the light intensity of the stars enough so that the nighttime would be dark.[2]

In the year 1823, Wilhelm Olbers,[3] for whom the paradox is now known, restated the problem and proposed, as had Chéseaux, an absorption mechanism, although a more modest one. According to Olbers' calculations, an amount of absorption of about one millionth of that proposed by Chéseaux should be sufficient. Later, Olbers' calculation was supported by the results of the famed Russian astronomer, F. G. W. Struve, in the year 1837.

Most astronomers evoked absorption to resolve the paradox of Olbers. They were forced to this explanation by their belief in an infinite universe. But in the twentieth century the paradox emerged once more, when evidence was discovered indicating an extreme transparency of space. Harlow Shapley, while studying distant globular clusters, reached the conclusion that little absorption could occur if his observations were correct. This research implied that intergalactic space was more transparent than some astronomers had

182

been willing to admit. (Unfortunately, Shapley incorrectly generalized his results to the interstellar regions in the plane of our Galaxy, which actually are strongly absorbing. See Section Two for details.) Consequently, Olbers' paradox demanded solutions other than those advanced by Chéseaux, Olbers, and Struve. Shapley, in the year 1917, reverted to a finite universe: "Either the extent of the star-populated space is finite or 'the heavens would be a blazing glory of light'. . . . Then, since the heavens are not a blazing glory, and since space absorption is of little moment throughout the distance concerned in our galactic system, it follows that the defined stellar system is finite."[4]

An ingenious method of circumventing Olbers' paradox was presented by C.V.I. Charlier in preliminary form in 1908 and in full detail in 1922.[5] His solution, which had originally been suggested by J. Lambert in 1761 but was not entertained seriously, involved a hierarchical grouping of stellar systems. The first-order system suggested was the grouping of stars into galaxies of a given diameter. Then a metagalaxy, i.e., a grouping of galaxies, formed the next order, which possessed a much larger diameter. The next order would be a grouping of metagalaxies, and so on. Charlier rigorously treated the mathematical requisites necessary to resolve Olbers' paradox. Charlier's proposal found several supporters because it was a simple solution, not only to the paradox noted by Chéseaux and Olbers, but also to an analogous problem within Newton's theory of gravitation, which will be discussed later.

Although Charlier attracted a few supporters, his ideas had little observational basis until Hubble demonstrated the equivalence of our Galaxy and spiral nebulae in 1924; furthermore, even though Hubble found evidence for clusters of galaxies, no evidence for higher orders of clustering was found during the 1920s. Consequently, the popularity of Charlier's hypothesis declined.

A resolution of Olbers' paradox acceptable to most astronomers awaited the cosmological observations of Hubble and the relativistic cosmological models of an expanding universe introduced by Friedmann, Lemaître, de Sitter, Einstein, and others. (These developments are discussed in Chapter Three.) The decrease in radiant energy from distant stars due to a red shift of the spectrum, which is a consequence of an expanding universe, is sufficient to avoid the paradox. Because the energy of the light from the farthest stars decreases the most, the summation of the contribu-

C.V.I. Charlier. (Lick Observatory photograph)

183

tions from all distances remains small compared to the light we receive from the Sun.

The Failure of Newtonian Mechanics

Like the intensity of light, the intensity of a gravitational field decreases as the square of the distance from its source. Through arguments parallel to those that led to Olbers' paradox, some persons noticed that the Newtonian formulation of gravitational mechanics encountered difficulties when applied to the entire cosmos.[6] Sir Isaac Newton came close to recognizing the gravitational equivalent to the optical dilemma but did not quite grasp the importance of the issue. Newton might never have considered the problem if the Reverend Richard Bentley, who was not a physicist, had not written to ask for clarification of some of the principles involved in the new theory of gravitation. Bentley, who was preparing a lecture to demonstrate the support science provided for religion—an increasingly difficult task—delved into the details of Newton's explanation of the dynamics of the solar system and the universe.

The correspondence between Bentley and Newton indicates that Bentley was concerned that a finite universe would be unstable and collapse towards the center—a view with which Newton concurred. The alternative—an infinite universe—also troubled Bentley. He assumed that the gravitational effects of all matter in the universe on one body should sum to zero. That is, all the gravitational attraction in one direction should sum to infinity; all in the other direction should also sum to infinity. The two opposite infinite pulls on the earth, for example, should cancel each other, and the earth should move in a straight line instead of around the Sun.

Newton responded that the two opposing infinities might not be equal. He also noted that the same analysis could be applied disregarding the presence of the sun; then, introduction of the solar mass would supply a gravitational force in addition to the two opposite and infinite forces. If the two opposing forces cancelled, then the solar force would pull the earth into its orbit.

Bentley also recognized that an infinite universe would be unstable. Unless there was perfect symmetry, evenly distributed particles would coalesce into larger lumps of matter. Newton agreed and argued that star formation would result. But Newton did not continue the argument to a coalescence of the stars.

Both men realized that the application of Newtonian gravitational theory to the universe was problematic. Although Newton side-stepped the issues with Bentley and published no comment on the difficulties, he had been forced to face a dilemma that he had previously ignored.

Newton's unpublished manuscripts indicate that he wrestled with the cosmological problems in private.[7] He concluded that a high degree of symmetry was required to prevent disruption of a static universe. To prove the uniformity of the universe, he attempted to deduce the stellar distribution from the number of stars at each stellar magnitude. He first assumed that the magnitudes indicated distances directly; 2nd magnitude stars were twice as far as 1st magnitude stars, 3rd magnitude stars were triple the distance of 1st magnitude ones, and so on. He soon discovered that the stellar magnitude data, coupled with his basic assumption on distances, indicated an increase in the number of stars with distance from the sun. To remove the difficulty, Newton revised his linear magnitude-distance assumption to one that would support a uniform universe. In other words, he choose a magnitude-distance relation so that the stellar distribution calculated from the stellar magnitudes would be uniform, and then used that relation to prove the uniformity of the universe—a very circular proof! But by so doing, Newton was able to remove the need for further consideration of the problems imposed by an infinite universe (at least as far as he was concerned). Indeed, few people did reconsider the problems.

In the year 1895, Hugo von Seeliger[8] presented a mathematical treatment of the same problems. In a rigorous fashion, he confirmed the suspicions of Bentley—a uniform, infinite universe was not stable according to Newtonian mechanics. Instead of resorting to circular reasoning as had Newton to circumvent the problem, von Seeliger modified the dynamical equations by an exponential factor that reduced gravitational effects for astronomical distances. Since the factor did not alter Newton's basic equations significantly for terrestrial or solar distances, its validity could not be tested locally. Not surprisingly, this conjecture was attacked for its *ad hoc* basis.

Hugo Seeliger. (Lick Observatory photograph)

Einstein lecturing in Pasadena, California, *c.* 1932. (B. Hoffmann and H. Dukas, *Albert Einstein: Creator and Rebel,* Viking: New York, 1972)

CHAPTER 2

Relativistic Cosmologies:

First Attempts

Einstein's Theory of Gravitation

One solution to the difficulties encountered in the Newtonian approach to cosmology was provided by the General Theory of Relativity, developed by Albert Einstein. The motivating desire to construct a symmetrical framework, within which to discuss physical problems, led Einstein to apply non-Euclidean geometry to the structure of space and time. As a consequence, some of the problems inherent in Newtonian cosmologies, based on Euclidean geometry, were resolved. Non-Euclidean geometry was essential for Einstein's formulation of general relativity, and hence relativistic cosmology.

The beginnings of the development of non-Euclidean geometry can be traced to as early as the eighteenth century. Several mathematicians, including the famous Karl Gauss, discovered that even if some of the axioms of Euclid were replaced by other statements, a consistent deductive geometry could still be constructed. For instance, the axiom that one and only one line parallel to another can be drawn through a point could be replaced by the axioms that many or no parallel lines could be drawn through a single point. The nineteenth-century mathematicians, N. Lobachevsky and J. Bolyai, derived many of the consequences of non-Euclidean geometries where many parallel lines could be constructed.[1] Another non-Euclidean geometry, in which no parallel lines were possible, was studied by B. Riemann. As the non-Euclidean field gained popularity, new mathematical techniques—for instance, tensor calculus—were applied to the analytical descriptions of the various geometries. Thus, Einstein was supplied with the basic tools necessary for the development of his new physical concepts.

During the years 1912-1914, Einstein worked with a Swiss-German friend—the mathematician Marcel Grossman—in applying non-Euclidean geometry to the concept of space-time. As a result, he was able to incorporate gravitation as a deformation of the space-time geometry in the vicinity of a massive body. Although gravitation was thereby simplified philosophically—gravitational forces were no longer a necessary concept—the mathematics was increased in complexity. Nevertheless, the new theory of gravity introduced the possibility of approaching the problems of cosmology anew. A result of these years of work were the so-called field equations of Einstein which expressed the general geometric structure of the universe.

Einstein's Static Cosmology

Einstein was well aware of the cosmological difficulties involved in the Newtonian approach to physics. To avoid problems at infinity, Einstein modified his original equations by inserting a term involving an arbitrary λ.[2] (This symbol became known as the "cosmological constant.") Mathematically, Einstein's insertion of the additional constant is quite legitimate and, in fact, for the purposes of maintaining the most general character of his equations, the λ term should be included. Some people have even wondered why

Einstein. (Courtesy of H. Landshaff, New York. Original in the Einstein Archives. First published in: B. Hoffmann and H. Dukas, *Albert Einstein: Creator and Rebel*, Viking: New York, 1972)

187

Einstein excluded the constant from his original formulation. The answer to that question probably lies in his desire to devise a theory exclusive of arbitrary constants that would be dependent upon empirical determination. Nevertheless, to maintain a static universe, the λ term had to be included.

Notice that Einstein must have believed strongly in a static universe; otherwise he would not have felt compelled to introduce an arbitrary aspect of his theory. In fact, he stated in 1917 that the term "is necessary only for the purpose of making a quasi-static distribution of matter. . . ."[3]

Although he had overcome one problem involved with a static universe, the solution of his equations at

Albert Einstein: 1879-1955

Einstein was born in Germany. His early schooling did not capture his interest; he disliked the strict regimentation. His father, an unsuccessful businessman who ran an electrochemical plant, eventually awakened his scientific interest by introducing Albert to a compass, and by supplying him with books.

After the business failed in Munich, the Einstein family moved to Milan to try their luck there. Albert remained behind to finish his schooling, but soon quit and passed the remainder of the year amusing himself in Italy. Persuaded to return to his studies at a college in Switzerland, he renewed his interest in science and completed his studies in physics and mathematics.

After graduation, Einstein did not find steady employment for several years, but was finally appointed an examiner in the Swiss Patent Office in Berne. His mind was well suited to the work and he was able to complete his assignments rapidly, leaving time for thinking about physical problems. During his seven years at Berne, he laid the foundation for much of his later work.

In 1905, Eistein received his doctorate from the University of Zurich for a dissertation on molecular dimensions. Recognition for his scientific abilities soon followed, and he was offered several academic positions. During the period after completing his dissertation, Einstein worked on the General Theory of Relativity at several universities. Finally, after accepting a position in Berlin, he completed the theory.

Although his scientific work advanced rapidly during the period of World War I, his refusal to support the German cause and his attempts to preserve an international spirit in science led to attacks from several groups. Between World Wars I and II, the verbal attacks by right-wing, anti-semitic extremists increased, as did threats of physical violence. Finally, as Hitler came to power, Einstein fled

Einstein as a child. (The Einstein Archives. Previously published in: B. Hoffmann and H. Dukas, *Albert Einstein: Creator and Rebel*, Viking: New York, 1972)

infinity still presented difficulties. The concept of a closed universe in curved space, which had been examined by Einstein's contemporaries, provided the solution to his difficulties merely by dismissing the infinite dimensions. The structure of space-time in a finite, unbounded, non-Euclidean geometry subject to his field equations, allowed the solution of a static cosmological model free from the embarrassments of some other models.

Curved Space-Time

What is meant by curved space? The consideration of the two-dimensional geometry on the surface of a sphere can provide some insight into the concepts of

The Einsteins in Japan, *c.* 1923. (B. Hoffmann and H. Dukas, *Albert Einstein: Creator and Rebel*, Viking: New York, 1972)

Einstein at about age 14. (The Einstein Archives. Previously published in: B. Hoffmann and H. Dukas, *Albert Einstein: Creator and Rebel*, Viking: New York, 1972)

Germany, eventually arriving in the United States where he remained for the rest of his life.

During the 1930s, Einstein renounced the pacifist stand he had held during World War I; he believed that the threat presented by Hitler could be eliminated only by force. Aware of the possibility that top German scientists, such as Heisenberg, might develop a nuclear fission bomb, Einstein encouraged President Roosevelt to initiate efforts that led to the Manhattan Project and development of the atomic bomb. After the bomb was used, and the war ended, Einstein once again worked for the abolition of war through the development of a world government.

Although he was active in politics, he gave first priority in his life to science. As he is reported to have said during a political discussion: "Yes, time has to be divided this way between politics and our equations. But our equations are more important to me, because politics is for the present, but an equation like that is something for eternity."

Niels Bohr and Albert Einstein. (Courtesy of M.J. Klein, Yale University. Previously published in: B. Hoffmann and H. Dukas, *Albert Einstein: Creator and Rebel*, Viking: New York, 1972)

curved space. A flat creature that lives on the surface of a sphere can make measurements of various geometric configurations. If he restricts himself to regions whose dimensions are small compared with the radius of the sphere, he will discover that Euclidean plane geometry is satisfactory for most practical purposes. But if he travels farther, he will soon notice that angles of a triangle sum to more than 180°, parallel lines intersect, and the circumference of a circle is not π times the diameter. Not only could a flat creature notice the deviations from Euclidean geometry, he could also construct a "curved" geometry to explain his results. All his actions could be completely independent of any knowledge of a third dimension.

What other strange features might our flat friend notice? If he walked (or moved in whatever way flat creatures do) in a straight line he would eventually arrive back at his starting point. If he surveyed the area of his "universe," or started to paint the surface bright orange, he would find that a finite number of gallons of paint would suffice to cover every possible point, yet he would never encounter a single boundary or edge. In cosmological terms, the flat creature lives in a finite but unbounded universe.

Einstein's closed universe is analogous. Like our flat friend who does not know about a third dimension and has no need for it as long as he has studied Riemann's geometry, we can envision a curved, unbounded space of finite volume without the necessity of a fourth spatial dimension. All we need is our non-Euclidean geometry.

The de Sitter Universe

Other types of geometry also satisfy Einstein's field equations. In the year 1917, W. de Sitter[4] discovered that under certain conditions an infinite hyperbolic space was possible. Such a space is analogous to the two-dimensional saddle-like figure that is infinite in extent. To maintain a static character, however, de Sitter's model had to postulate that no matter existed, i.e., the average density of space is zero. Obviously, de Sitter's conditions cannot be satisfied exactly, but this solution could have approximate validity if the density were small enough.

De Sitter's cosmological model was a source of consternation to Einstein. Although the model provided a valid solution to his field equations, he was concerned about the implication of a massless universe—that a coordinate system could exist independent of a mass, i.e., an absolute property of space existed.

In spite of its massless characteristic, the de Sitter universe had some consequences of great astronomical interest. For instance, a result of the mathematical model was that atoms at great distances would emit light at lower frequencies, since their time would seem to slow down. Consequently, spectral lines from dis-

Visit of Albert Einstein to Yerkes Observatory, 6 May 1921. (Yerkes Observatory photograph)

Dr. and Mrs. DeSitter. (Sky Publishing Co. photograph)

tant parts of the universe would appear to be red-shifted. Furthermore, if test particles (small masses) were inserted into the universe, they would accelerate away from each other because of the λ term. Arthur S. Eddington, the famous English astronomer, recognized the redshift aspects of de Sitter's model and emphasized them in the early 1920s. Hence, de Sitter's model received wide publicity and evoked much interest among astronomers.[5]

Problems with the Einstein and de Sitter Cosmologies

The influence of astronomical observations—in particular, the research on the structure of our Galaxy and the status of the island-universe theory—was very important to the development of early relativistic, cosmological models. During the decades before 1917, considerable evidence had accumulated to indicate that the material distribution within our universe was confined to a relatively small volume. The Kapteyn universe restricted our local stellar system to a region whose dimensions were approximately 10 kiloparsecs. The Shapley galactic model, which was proposed about the same time that Einstein and de Sitter published their cosmological models, increased the possible dimensions by a factor of ten but certainly not to any extent approaching infinity. Although the de Sitter model did not require large distribution of matter, i.e., of material in the universe, the Einstein model did. Consequently, if the entire matter in the universe were localized to within the vicinity of our Milky Way Galaxy, as many astronomers believed before 1924, then the validity of the Einstein model was questionable. A possible means of salvaging his model was afforded by the island-universe theory, but few astronomers were

The "Relativists" gather at Leiden, 26 September 1923. (l. to r.) Einstein, Eddington, Ehrenfest, Lorentz, DeSitter. (Yale University photograph)

seriously considering that hypothesis when the relativistic models were presented. Einstein himself apparently did not believe that spiral nebulae were comparable to our Galaxy until this was demonstrated by Hubble in the mid-1920s (see Section Three). Clearly, Einstein's indifference to the incompatibility of his model with the prevailing astronomical theories of cosmical material distribution indicates that he was more interested in the beauty of a symmetrical idealization than in observational confirmation.

Willem de Sitter: 1872–1935

A drawing of DeSitter. (Yale University photograph)

Willem de Sitter was born in Sneek, Friesland (a northern province of Holland). After completing his preliminary education at the gymnasium in Arnheim, he enrolled in the University at Groningen. Although he intended to study mathematics, he was irresistibly drawn to astronomy. The year 1896 marked one of the most important events in his life. He met Sir David Gill, the director of the Royal Observatory at the Cape of Good Hope, who was so impressed with de Sitter's abilities as a measurer of photographic plates that he invited de Sitter to work with him at the Cape.

De Sitter waited until 1897 in order to finish his doctoral examinations and then joined Gill at the Cape, where he remained for two years. There he worked on a photographic photometric program to study the differences in colors of stars near the Milky Way plane and those near the galactic poles. He also began a project that became his passion throughout his life—a study of the satellites of Jupiter. De Sitter worked on this theme for more than 30 years, publishing some 30 major papers on the subject.

He returned to Groningen in 1899 and remained there until 1908, when he was appointed a professor of astronomy at the University in Leiden. Although his major interest remained in celestial mechanics, he published several papers on the theory of relativity beginning in 1911. De Sitter was one of the first scientists to appreciate the new theory. He worked out the astronomical consequences of Einstein's general theory in the period 1916–1917, and developed what has become known as the "de Sitter Universe"—a cosmological model that predicted redshifts of lines in the spectra of distant objects.

In 1919, he was appointed director of the observatory at Leiden and worked closely with Kapteyn to make the institution one of the foremost in the world. His abilities as an administrator also won him the presidency of the International Astronomical Union, a post he held from 1925 to 1928.

De Sitter received many honors, including the Gold Medals of the Royal Astronomical Society, the Astronomical Society of the Pacific, and the American National Academy of Sciences.

The galaxy NGC 4884 receding from us with a velocity of 7000 miles/sec. A 1930 photograph taken with Mount Wilson's 60-inch telescope. (Hale Observatories photograph)

CHAPTER 3

The Expanding Universe

The Demise of Static Models

The year 1917 marked the real beginning of modern quantitative cosmology, with the publication of the models of Einstein and of de Sitter. These models, based on the first applications of general relativity to cosmology, were remarkably different. Einstein's model[1] (called solution A) described a universe containing matter but static; de Sitter's model[2] (called solution B) described a universe devoid of matter but allowing motion (i.e., if two test particles were inserted they would appear to recede from one another).

Both of these models, however, were based on the same field equations, which included the cosmological constant term λ. This term, originally introduced by Einstein, was equivalent to a repulsive force, which was needed in order to overcome the difficulties of applying Newtonian gravitation to cosmology.

Both models (A and B) are termed "static," since even though B admits motion, there is no matter present in the universe to move. In order to differentiate between these models, two questions need to be answered by the observations: (a) Is the universe empty, i.e., does the mean density approach zero? (b) Is there motion in the universe?

194

De Sitter made the first test himself in 1930 after Oort had estimated the mass of the Galaxy as 10^{11} solar masses.[3] He found the universe far too full of matter to admit solution B.[4]

Since the motions of nebulae were (by then) well established (from the work of Slipher and Hubble), solution A was clearly invalid. De Sitter therefore concluded that "we thus come to the conclusion that both the solutions (A) and (B) must be rejected, and as these are the only static solutions of the equations, the true solution represented in nature must be a dynamical solution."[5]

The first alternative to the static solutions had been proposed by Aleksandr Friedmann in 1922. He introduced a non-static solution to Einstein's relativistic equations; Friedmann assumed that the universe could evolve with time. His solutions avoided some of the problems inherent in the static models.

Independently, though several years later, the Abbé George Lemaître arrived at similar conclusions and published a non-static model. Unfortunately, the works of these two men did not attract the attention of astronomers until 1930.

Observations

The observational evidence, which was to have a profound effect on the development of cosmology, began to accumulate as early as 1914, with the pioneering work of Slipher at the Lowell Observatory.

The results of Slipher's research showed that almost all spiral nebulae are moving away from our Galaxy (and also away from each other) with very large velocities. Not only did this fact hint that the spirals were galaxies like our own (as was discussed in Section Three) but it also suggested to several astronomers that a systematic motion of the universe was being observed. For example, G.F. Paddock wrote in 1916 that "the average velocity . . . is decisively positive, which means that they are receding not only from the observer or star system but from each other. Accordingly, a solution for the motion of the observer through space should doubtless contain a constant term to represent the expanding or systematic component whether there be actual expansion or a term in the spectroscopic line displacements not due to velocities."[6]

The reason astronomers began investigating a systematic motion was the following: If the motions of the nebulae are assumed to be random, the average of the values should reveal the motion of the observer

195

(i.e., the Sun) through our own star system. As information accumulated, however, it was found that the nebular velocities could be accounted for only if a systematic term was present.

Using his early data, Slipher found that the nebulae on one side of the Milky Way star system had velocities greater than those on the other side. From this he postulated that our star system was drifting with respect to the nebulae. With more measurements, however, this effect disappeared—almost all nebulae were moving away from the Milky Way.

In 1918, the German astronomer C.W. Wirtz attempted to calculate the motion of the Sun with respect to the spiral nebulae. He initially assumed that the motions of the spirals were random. He found, however, that he always had a large residual radial velocity term. He then introduced a term into the calculations to account for the systematic effect.[7] As Slipher continued to supply data, Wirtz and Lundmark calculated the residual term to be about 800 km/sec. Wirtz went further and in 1922 suggested that the term might represent a systematic recession of spiral nebulae from the Sun.[8]

Edwin Hubble in front of the Hooker telescope in 1952. (Hale Observatories photograph)

The next step taken by Wirtz appears to have been influenced by a theoretical development made by de Sitter in 1917.[9] De Sitter had stated that in one case his solutions of Einstein's equations indicated a systematic displacement of spectral lines for distant nebulae.

Wirtz investigated the possibility that the residual radial velocity term might be a function of distance.[10] Since distance measurements of the spirals were not available at this time, he tried to correlate the term with the apparent diameters but was unsuccessful. K. Lundmark and B. Stomberg also tried to make such a correlation, but were equally unsuccessful. Stromberg found no correlation and Lundmark found a very weak one. The problem was that, again, reliable distance measurements were needed.

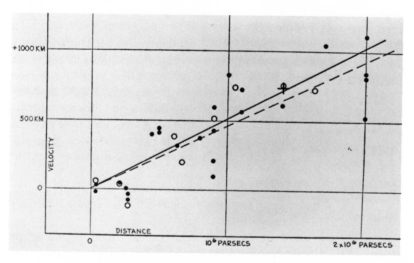

After 1925, when Hubble discovered Cepheids in spirals, distances could be found and the radial velocity as a function of distance could be investigated more carefully. By 1929, although the velocities of 46 objects were available, Hubble had only been able to obtain distances for 18. Nonetheless, he presented a paper before the National Academy of Sciences that year that seemed to demonstrate a clear relationship between velocity and distance for distant spiral nebulae.[11] This relation has now come to be known as the "Hubble law."

This relation between velocity v and distance d,

$$v = H\,d$$

(the constant H is the modern usage; Hubble used K), described the linear increase of recessional velocity with increasing distance.

Hubble's velocity-distance relation. (E.P. Hubble, *Realm of the Nebulae*, Yale University Press: New Haven. 1936)

197

Milton Humason measuring spectra for red-shifts.

After 1929 Hubble, working with M. Humason, began a program to extend the observations of velocities and distances in order to verify the velocity-distance relation. They published a paper is 1931[12] that extended Hubble's earlier results. In discussing this paper, Hubble stated that "the velocities over the entire range increase directly with distances, and the linear relation holds as closely as distances can be estimated."[13]

By 1934, they had distances to 32 individual nebulae and over a hundred velocities.[14] The relation seemed to be firmly established.

Soon after Hubble's publication, many cosmological theories concerning an evolving universe were discussed. Most astronomers and cosmologists interpreted Hubble's law as the observational evidence for an expanding universe.

Theoretical Interactions

The theory of de Sitter—which predicted that redshifts of spectral lines should increase with distance—was published in 1917. Since Slipher's results showing large recessional velocities for spiral nebulae were available at this time, it is interesting that de Sitter did not use the data to attempt to verify his work. Also, Wirtz, who did try to use Slipher's results to try to relate distance to velocity, did not use de Sitter's results until the mid-twenties. One possible reason for this lack of communication has been described by Hubble: "Slipher's list of 13 velocities, although published in 1914, had not reached de Sitter, probably as a result of the disruption of communications during the war. For the same reason, Wirtz, in 1918 was probably not aware of de Sitter's papers."[15]

Exactly what effect the war had on scientific communications is not yet fully known. It is not impossible that the various investigators did have knowledge of each others' work. The important point, however, is that reliable distances were necessary before any correlation with de Sitter's work was possible.

As soon as Hubble had compiled enough distances, he presented a velocity-distance relation. The cosmological work of de Sitter obviously influenced him, as is evidenced by a comment at the end of his 1929 paper: "The outstanding feature, however, is the possibility that the velocity-distance relation may represent the de Sitter effect, and hence, that numerical data may be introduced into discussions of the general curvature of space."[16]

As previously mentioned, although several other astronomers were investigating the relationship between velocity and distance, all were unsuccessful—it remained for Hubble to achieve in 1929. The development of the relation was confused because of the work of Ludvik Silberstein, in 1925.[17] He tried to verify de Sitter's predicted relation, using globular clusters whose distances were better determined than those of the spirals (although the clusters did not show the systematic redshifts, as did the spirals). De Sitter's effect, however, was derived in terms of apparent Doppler effects due to curvature of space, which yielded only positive (i.e., recessional) velocities. Since

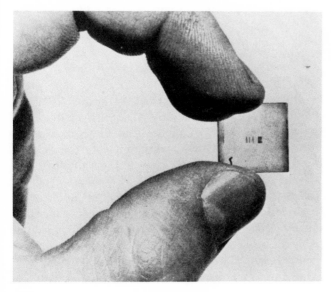

A typical spectrogram. The plate measures 15 mm x 15 mm. (A. Sandage, "The Red Shift," *Scientific American*, Sept. 1956)

Arthur Stanley Eddington. (Yerkes Observatory photograph)

the Andromeda Nebula showed a negative velocity, Silberstein thought de Sitter's prediction was contradicted.

He derived a velocity-distance relation in a manner similar to de Sitter's approach (but invoked somewhat controversial assumptions) and found a result that yielded both positive and negative velocities. Application of his analysis to some data for local features (globular clusters and the Andromeda nebula) indicated agreement. But astronomers soon refuted his thesis with a more thorough examination of the data.

The effect of all this was to create a climate of doubt for any further work in the area. Because some workers might link his work to earlier attempts, Hubble was careful in 1929 to distinguish between observation and theory. He emphasized the observational nature of his work and mentioned de Sitter only once, briefly and in the last paragraph of his paper.

Hubble's work had a powerful effect on Einstein. His original cosmology called for a static universe, that is, one that is neither expanding nor contracting. In 1930, however, he was forced to give up the model, and wrote that "new observations by Hubble and Humason concerning the red shift of light in distant nebulas [sic] make it appear likely that the general structure of the universe is not static."[18] Einstein travelled to Mount Wilson to confer with Hubble, and early in February 1930 announced that he had given up his static model of the universe.

While on Mount Wilson, Einstein and his wife Elsa were given a tour of the observatory. It was explained to them that the giant telescope was used for determining the structure of the universe, to which Elsa replied, "Well, well, my husband does that on the back of an old envelope."

Expanding Models

By 1930 it was obvious that static models of the universe were inadequate. At a meeting of the Royal Astronomical Society in January of that year Eddington noted: "One puzzling question is why there should be only two solutions. I suppose the trouble is that people look for static solutions."[19] Upon reading this remark, Lemaître, a former student of Eddington's, informed Eddington that he had published a nonstatic model of the universe in 1927. Eddington then saw to it that Lemaître's paper was published in the *Monthly Notices*, where it would receive a wider circulation.[20] Thus the expanding universe was born.

As mentioned earlier, this was not the first nonstatic solution; Friedmann had published solutions in 1922[21] and 1924.[22] Apparently the importance of these papers (and of Lemaître's) had escaped the attention of astronomers until Eddington's above remark. The main difference between the models of Friedmann and Lemaître involved the use of the cosmological constant —Lemaître employed the constant and Friedmann did not. The models were similar, however, in that both were based on solutions of Einstein's relativistic field equations, and both described a universe in expansion.

A strong item of evidence in favor of the nonstatic solutions was Hubble's observed relation between velocity and distance. Although this relation by itself did not lead to the downfall of the static solutions (the observed mean density of the universe played a vital role), it did strengthen belief in the nonstatic model since it appeared to constitute observational verification of them.

The first formulation of the relation, however, presented a problem to the expanding models. The inverse of the constant H expresses the age of expansion of the universe. For the models that described the expansion from a super-dense point, $1/H$ represented the age of the universe. Hubble originally found $H = 500$ km/sec Mpc,$^{-1}$ corresponding to an age of the universe of 10^9 years. This cosmological age was embarassingly small when compared with the independently determined ages of the stars by Arthur S. Eddington (in 1924)[23] and James Jeans (in 1928).[24] Eddington had shown that the relation between stellar

Observing at Mount Wilson: (*l. to r.*) Einstein, Hubble, Adams. (Henry E. Huntington Library)

(*l. to r.*) Walter Baade, Edwin Hubble, Abbé LeMaître at the observers' residence (called "the monastery") on Mount Wilson, *c.* 1932. (Henry E. Huntington Library)

The Abbé George LeMaître. (Yerkes Observatory photograph)

George Lemaître: 1894-1966

George Lemaître was born in Charleroi, Belgium, and received his early education in the Jesuit school there. He enrolled at the University of Louvain in 1911 as an engineering student, but had to terminate his studies in 1914 because of the war. He served gallantly in the Belgian army, winning the Croix de Guerre. He returned to the University of Louvain in 1918 but switched to studying mathematics and physics and obtained his doctorate in 1920. He immediately entered the seminary at Mailines, and was ordained a Roman Catholic priest in 1923.

Lemaître won a scholarship from the Belgian government that enabled him to travel and study abroad. He used this to great advantage by spending a year at Cambridge University, where he studied with Eddington, and another year at the Harvard College Observatory and the Massachusetts Institute of Technology.

He returned to the University of Louvain in 1925 where he remained for the rest of his life, achieving the rank of full professor in 1927.

Lemaître was interested in many different areas of research, including astrophysics, cosmic rays, celestial mechanics, and numerical methods. He is best known, however, for his work in cosmology. In 1927 he proposed an expanding relativistic model of the universe, which was overlooked by other scientists until Eddington drew attention to Lemaître's paper by having it reprinted in the *Monthly Notices*, in 1930. Lemaître also championed the concept of the "primeval atom," beginning in 1945.

He received many awards during his life, including the Prix Franqui in 1934 and the Royal Society's Eddington Medal in 1953 (Lemaître was the first recipient). He was elected to the Royal Society (1939), the Pontifical Academy of sciences (1940, serving as president from 1960), the Belgian Academy of Sciences (1941), the American Philosophical Society (1945), the Academi Nazionale Dei XL (1961), and the Academia Internationale 'Neocastrum' (1966).

masses and luminosity could be explained if the stellar material obeyed the perfect gas law and energy production were due to annihilation of electrons and protons. The time required to radiate one solar mass was estimated to be about 2×10^{13} years. Jeans applied the theory of relaxation times in a gravitating ideal gas to groups of gravitationally interacting stars and found that the partial exchange of energies derived from statistical analysis of stellar positions and motions indicated a time on the order of 10^{13} years. The close correspondence of the results of Eddington and Jeans made the value for the ages of the stars seem very believable.

The age of the earth was calculated around 1928 from the relative abundance of radioactive elements and their decay products. The calculation indicated an age of the earth of about 2 to 6×10^9 years. Thus, these values placed the birth of the stars and of the earth before the beginning of the universe!

The difficulties were reconciled later when better theories of stellar dynamics and stellar evolution were formulated and when better estimates of the distances to galaxies (and hence H) became available. The accepted value of H today is only about one tenth of that originally given by Hubble. Discoveries of fundamental differences between classes of stars and refinements in distance estimation techniques have steadily increased the calculated age of the universe.

As noted earlier, models of an expanding universe imply a finite cosmological lifetime. From the present observed rate of expansion astronomers can theoretically follow the evolution of the universe back in time until all matter was compacted into a small region. Since the present universe seems to have evolved from a condensed state (often called the "primeval atom") through some explosion (to provide the expansion mechanism), the "birth" of our universe is often referred to as the "big bang" and corresponding cosmological models are known as "big bang" cosmologies.

Another cosmological model was introduced in 1948 by Hermann Bondi and Thomas Gold.[25] This model postulated an expanding universe with no beginning and no end. The theory—known as the steady-state theory—maintained that the universe is infinitely old and has always appeared as it does now. There is a problem, however, in that the cause of the expansion is not specified in contrast to the big-bang models that account for the expansion as a result of an explosion at the initial point.

Newspaper article on the expansion of the universe. (*Boston Globe*, 22 November 1931)

203

One of the most fundamental differences between the two types of theories (big-bang cosmologies and the steady-state cosmology) involves the concept of creation. The big-bang models insist that all the matter in the universe was created at once—at the beginning. The steady-state model insists that matter is being created continuously in order to preserve a constant density.

This continuous creation is fundamentally inconsistent with the existing picture of physics in that the basic principles of conservation are violated. If matter is created continuously, the amount of matter is never a fixed quantity. Hence, the laws of conservation of mass and of energy no longer hold. Admittedly there is nothing sacred about conservation laws. But, since they play such a fundamental role in our understanding of physics, it is difficult to discard them unless absolutely forced to.

These two cosmological theories provoked a great deal of debate and inspired much of the research in cosmology during the 1950's and 1960's.

The cover of *Time* magazine for 9 February 1948 showing a portrait of Edwin Hubble. (Reprinted by permission from *Time*, the Weekly News magazine; Copyright Time Inc.

Hubble using the 48-inch Schmidt camera at Mount Palomar. (Hale Observatories photograph)

CHAPTER 4

Interpretation of Redshifts

Redshifts as Velocities

Hubble's discovery in 1929 of the relation between redshift and magnitude for nebulae (the velocity-distance relation) had a powerful impact on cosmology. As Hubble himself noted, "The velocity-distance relation is not merely a powerful aid to research, it is also a general characteristic of our sample of the universe —one of the very few that are known . . . if it could be fully interpreted, the relation would probably contribute an essential clew to the problem of the structure of the universe."[1] His phrase, "if it could be fully interpreted," reflected the outstanding difficulty with the relation that Hubble and many others found to be a major obstacle.

205

Edwin Powell Hubble: 1889-1953

Edwin Hubble ready to enjoy his favorite hobby, fly-fishing. (E.P. Hubble, *The Nature of Science, and Other Lectures*, Henry E. Huntington Library: San Marino, California, 1954)

Hubble, like Shapley, was born in Missouri and arrived at a career in astronomy through a somewhat roundabout fashion.

He received a B.S. degree in astronomy from the University of Chicago in 1910 and was awarded a Rhodes Scholarship to study law at Oxford. An active athlete throughout his life, he took part in track, boxing (at which he was successful enough to be considered as a contender for the heavyweight championship), and rowed stroke for the Queen's boat. Returning to the United States in 1913, he passed the bar examination and half-heartedly practiced law for a year in Kentucky. Fortunately, he returned to science because he felt that "even if I were second-rate or third-rate, it was astronomy that mattered." Thus, in 1914, he began graduate work at the University of Chicago. He finished his Ph.D. work in 1917 and, although invited to join the Mount Wilson staff, joined the army instead. He telegraphed Hale, "Regret cannot accept your invitation. Am off to war." He achieved the rank of major and was wounded in France. Again in 1942, Hubble set his research aside to do war work in ballistics at the Aberdeen Proving Ground.

An accounting of Hubble's research is essentially the history of modern extragalactic astronomy. One of his most important achievements, the discovery of Cepheids and other variable stars in nebulae, resulted in establishing the size and structure of the universe. This, in turn, served as the beginning of investigations concerning the evolution of the basic structures of the universe and the founding of the velocity-distance relationship, known as Hubble's law. These results brought about the birth of modern cosmology. Altogether, they undoubtedly rank as the most significant contributions to cosmology since the time of Copernicus.

The relation was an observational one between magnitudes of nebulae and their redshifts. Although, astronomers were fairly confident that distances could be estimated from magnitudes, the interpretation of redshifts did not inspire such confidence. Many people interpreted the redshifts simply as velocities from the Doppler principle, i.e., for a shift $\Delta\lambda$ of any particular wavelength λ, the velocity v is found from the relation

$$c \, \frac{\Delta\lambda}{\lambda} = v$$

where c = velocity of light. The redshifts indicated velocities of recession, implying an expanding universe.

Hubble felt that the Doppler interpretation of redshifts was not completely obvious and insisted on using the wording "apparent velocities" to indicate this:

206

Hubble (*center*) posing with several movie stars at the M.G.M. studios in Hollywood. The silent screen star Raymond Navarro is immediately to Hubble's left. (Henry E. Huntington Library)

Hubble's classic book, *Realm of the Nebulae* (1936), is still a source of inspiration as well as a basic text on astronomy today. Sir James Jeans reviewed the book (the first review he had consented to write in 10 years) and wrote that it was "a chapter of scientific history which has stirred the imagination not only of professional astronomers but also of the public at large."

Hubble also wrote and gave lectures on a wide variety of topics, including the exploration of space, the role of science in liberal education, science and citizenship, Francis Bacon, English science in the Renaissance, smog, surveys with the 48-inch Schmidt telescope, and a BBC program dealing with the 200-inch telescope. He received many awards, including election to the National Academy of Sciences, five honorary degrees, and numerous gold and silver medals.

"Meanwhile redshifts may be expressed on a scale of velocities as a matter of convenience. They behave as velocity-shifts behave and they are very simply represented on the same familiar scale, regardless of the ultimate interpretation. The term 'apparent velocity' may be used in carefully considered statements, and the adjective always implied where it is omitted in general usage."[2]

He always remained cautious when discussing the redshifts. In 1931 he wrote to de Sitter, "We use the term 'apparent' velocities in order to emphasize the empirical features of the correlation. The interpretation, we feel, should be left to you and the very few others who are competent to discuss the matter with authority."[3]

Apparently Hubble was not convinced that the concept of the expanding universe was a valid one. N. Mayall wrote to him from Lick in 1937: "It is perhaps unnecessary to mention how pleased are some of the

Hubble shows Dr. Richard Tolman, noted mathematical physicist, a model of the proposed 200-inch telescope, May 1931. (Henry E. Huntington Library)

people here to note the way that your interpretation of the nature of the redshift casts doubt upon the validity of theories of the expanding universe."[4]

Other Explanations

Although many people eagerly accepted the interpretation of redshifts as velocities, some (most notably Hubble) felt that other interpretations were also possible.

Hubble commented on the view held by most theoreticians:

The interpretation as velocity-shifts is generally adopted by theoretical investigators, and the velocity-distance relation is considered as the observational basis for theories of an expanding universe. Such theories are widely current. They represent solutions of the cosmological equation, which follow from the assumption of a nonstatic universe. They supersede the earlier solutions made upon the assumption of a static universe, which are now regarded as special cases in the general theory.[5]

He continued with the view *he* held as an observer: "Because the telescopic resources are not yet exhausted, judgment may be suspended until it is known from observations whether or not redshifts do actually represent motion."[6]

Hubble insisted on suspending judgment because other interpretations of the redshift were possible. Given the basic relation between energy and wavelength for radiation proposed by Einstein in 1905,

$$\text{Energy x wavelength} = \text{constant,}$$

an increase in the wavelength for distant nebulae (i.e.,

208

a redshift) would indicate a loss of energy in the radiation. This loss could occur either in the nebula itself (the Doppler effect) or in the long path through space that the light must travel.

Hubble felt that observational tests could be made to distinguish between these two views. He argued that rapidly receding nebulae should appear fainter than stationary nebulae at the same distances. Thus, all that need be done, in principle, was to determine a scale of absolute magnitudes of nebulae and then compare what their apparent magnitudes should be at various distances. Then, by comparing the apparent magnitudes of nebulae with redshifts in their spectra against the computed values of what their magnitudes would be if stationary, the issue could be clearly decided.

Although this procedure is simple in principle, it is extremely difficult in actual practice. The individual differences among nebulae and the difficulty of making the precise measurements necessary for objects that are extremely distant make the test nearly impossible to perform. Hubble was never able to obtain the results necessary to decide between the different interpretations of redshifts.

At the Athenaeum, 1931. Seated (*l. to r.*) are Millikan, Einstein, Michelson, Campbell; standing: St. John, Mayer, Hubble, Munro, Tolman, Balch, Adams, Ballard. (Henry E. Huntington Library)

Although most scientists today readily accept the redshifts as velocities, the matter is not yet completely settled. Many still argue for other interpretations and insist that the velocity-distance relation may not represent an expanding universe.

Epilogue

for Section Four

Cosmology became an important topic of astronomical discussion as a result of the introduction of relativistic mechanics and its application to the structure of space. Two separate cosmologies were devised in 1917 that satisfied the restrictions imposed by Einstein's field equations; each one was based on the assumption that the universe is static. The Einstein model proposed a closed universe with a smooth distribution of matter, and the average density was assumed to be small but finite. The de Sitter model postulated that space was essentially empty.

Certain problems with the static models became apparent in the late 1920s. The discovery of many extragalactic nebulae displaying large spectral redshifts caused serious concern about the validity of Einstein's solution. Although the redshifts were predicted by de Sitter's model, the average density of the universe was determined to be too large to satisfy the approximation of zero-density required.

In 1930, some scientists realized that perhaps the static assumption was the cause of their cosmological troubles and directed their attention toward possible nonstatic solutions. Unknown to most astronomers, several nonstatic cosmologies had already been devised. These models, however, were brought to light only when observation provided a compelling need to consider theories describing an expanding universe.

Modern observational cosmology began in 1929, when Hubble's velocity-distance relation provided a basis for discussing the nature of the universe as a whole. The research of the early 1930s into the distribution of matter in the universe, coupled with Hubble's relation, forced the demise of the static models.

The arguments over the real nature of the observed redshifts, however, have never been settled. Nor have the arguments been quieted over which of the newer

211

cosmological models (if any) is an accurate picture of the universe. Some of the dilemmas encountered in the 1930s have been resolved, but others have resisted the efforts of astronomers and remain on the frontier of current research.

One problem that has been very important in recent cosmological research concerns whether the big-bang theory or the steady-state theory is a correct model for the universe. In 1965, a new discovery was made that had a profound effect on these cosmological models. It was found that the entire universe is apparently filled with a background radiation of 3°K. This radiation was, in fact, predicted in 1948 as a consequence of the big-bang explosion. It was argued that a remnant of the radiation from the hot matter of the initial explosion should have remained as background radiation. This radiation would be diluted (i.e., cooled) as the universe expanded so that, at present, it would appear to be about 3°K. This is perhaps one of the most convincing arguments in favor of the big-bang theory.

Although the steady-state theory has by now been rejected by almost all astronomers, considerable controversy remains concerning the nature of the evolutionary universe associated with the big bang. Did the primordial universe explode with enough energy to keep expanding forever? Or will it, like a rocket

---

Sample calculation:
Red Shifts and Radial Velocities

If a certain spectral line were known to occur at a specific wavelength, λ_L, (i.e., measured in the laboratory where $v=0$), but the line was observed to occur at a different wavelength, λ_O, when seen in the spectrum of an astronomical object, the radial velocity of the object could be computed from the relation:

$$\frac{\Delta\lambda}{\lambda} = \frac{v}{c}$$

where

$\lambda = \lambda_L$

$\Delta\lambda = \lambda_O - \lambda_L$ (amount the line is shifted)

$c =$ velocity of light $= 3 \times 10^5$ km/sec

$v =$ radial velocity

For example, if $\lambda_L = 4000$ Å and $\lambda_O = 4013$ Å, then

$$\frac{\Delta\lambda}{\lambda} = \frac{4013\text{-}4000}{4000} = \frac{13}{4000} = \frac{v}{c}$$

$$v = \frac{13}{4000} \times 3 \times 10^5 = 1000 \text{ km/sec.}$$

shot up with less than escape velocity, eventually slow to a stop and then fall back again?

A universe that continues to expand is called "open" and its curved space is the kind described by Lobachevsky. A universe that ultimately contracts and collapses is called "closed" and its space of the spherical, Riemanian sort. How can these be distinguished? The redshifts of distant galaxies, representing a much earlier epoch in the history of the universe, show a deviation from a uniform rate of expansion—the outward rush of galaxies is slowing because of their mutual gravitational attractions. If there is enough matter in the universe (that is, if the density is high enough), the expansion should be retarded to produce a closed universe. With less matter (a density below about 10^{-28} gm/cm^3) the expansion would be insufficiently slowed to close the universe. In principle the redshifts

A cluster of galaxies in the constellation Hercules photographed by the 200-inch telescope at Mount Palomar. (Hale Observatories photograph)

could distinguish between these two models, and much effort has been expended in this attempt. So far the results are inconclusive because the distance determinations of the farthest galaxies are too ambiguous to settle such a delicate point.

Many astronomers now hope that measurements of the density of the universe can settle the controversy as to whether the universe is open or closed. Here a curious situation exists. The measurements seem to show that the visible material in the universe has a density of approximately 10^{-30} gm/cm^3, many times too small for a closed universe. Yet many astronomers are reluctant to accept this solution. Is there some "missing mass," for example, in the form of massive black holes or in enormous outlying but overlooked regions of galaxies? If so, perhaps the universe is closed after all.

There is still controversy in astronomy—just as in the days of arguments over the island-universe theory. The wheel has apparently gone a full turn between 1920 and 1970, and even though the scale of the problem has expanded from galactic to cosmic, people today are again living in an era of uncertainty over the nature of their universe.

References

Chapter 1

1 S. L. Jaki, *The Paradox of Olbers' Paradox* (New York: Herder and Herder, 1969): 38.
2 *Ibid.*, 78.
3 *Ibid.*, 89–90.
4 H. Shapley, "Studies Based on the Colors and Magnitudes in Stellar Clusters, Part II," *Astrophys. J.* 45 (1917): 139.
5 J. D. North, *The Measure of the Universe* (London: Oxford University Press, 1965): 20–22.
6 Jaki, *Paradox*, 60 ff.
7 M. Hoskin, *The Listener* (30 July 1970). See also, O. Gingerich, "Astronomy Three Hundred Years Ago," *Nature* 225 (1975): 602–606.

Chapter 2

1 E. Lukas, "Non-Euclidean Geometry"(in press).
2 J. D. North, *The Measure of the Universe* (London: Oxford University Press, 1965): 56.
3 A. Einstein, "Kosmologische Betrachtungen zur Allgemeinen Relativitätstheorie," *Sitzungs-berichte der Preussischen Akad. d. Wissenschaften* (1917); English translation by Lorentz, *The Principle of Relativity* (New York: Dover, 1952): 177–188.
4 W. de Sitter, "On Einstein's Theory of Gravitation and Its Astronomical Consequences," *Mon. Not. R. Astron. Soc.* 78 (1917): 3.
5 North, *Measure*, 95–96.

Chapter 3

1 A. Einstein, "Kosmologische Betrachtungen zur Allgemeinen Relativitätstheorie," *Sitzungs-Berichte der Preussischen Akad. der Wissenschaften* (1917); 142, English translation by Lorentz, *The Principle of Relativity* (New York: Dover, 1952).
2 W. de Sitter, "On Einstein's Theory of Gravitation and Its Astronomical Consequences," *Mon. Not. R. Astron. Soc.* 78 (1917): 3–28.
3 J. Oort, "Investigations Concerning the Rotational Motion of the Galactic System," *Bull Astron. Inst. Neth.* 4 (1927): 79–89.
4 W. de Sitter, "On the Distances and Radial Velocities of Extra-Galactic Nebulae and the Explanation of the Latter by the Relativity Theory of Inertia," *Proc. Nat. Acad. Sci. USA* 16 (1930): 474–488.
5 *Ibid.*, 474.
6 G. F. Paddock, "The Relation of the System of Stars to the Spiral Nebulae," *Publ. Astron. Soc. Pac.* 28 (1916): 109.
7 C. W. Wirtz, "Uber die Bewegungen der Nebelflecke," *Astron. Nachr.* 206 (1918): 109.
8 C. W. Wirtz, "Notiz. sur Radialbewegungen der Spiralnebel," *Astron. Nachr.* 216 (1922): 451.
9 W. de Sitter, "On Einstein's Theory," 3–28.

10 C. W. Wirtz, "De Sitter's Kosmologie und die Bewegungen der Spiralnebel," *Astron. Nachr.* 222 (1924): 21.
11 E. Hubble, "A Relation Between Distance and Radial Velocity Among Extra-Galactic Nebulae," *Proc. Nat. Acad. Sci. USA* 15 (1929): 168–173.
12 E. Hubble and M. Humason, "The Velocity-Distance Relation Among Extra-Galactic Nebulae," *Astrophys. J.* 74 (1931): 43.
13 E. Hubble, *Realm of the Nebulae* (New Haven: Yale University Press, 1936): 117.
14 E. Hubble and M. Humason, "The Velocity-Distance Relation for Isolated Extra-Galactic Nebulae," *Proc. Nat. Acad. Sci. USA* 20 (1934): 264.
15 Hubble, *Realm*, 109.
16 Hubble, "Relation Between Distance and Radial Velocity," 173.
17 For a more thorough discussion of Siberstein, see: N. S. Hetherington, *The Development and Early Application of the Velocity-Distance Relation* (Ph.D. Dissertation, Indiana University, 1970).
18 R. W. Clark, *Einstein: The Life and Times* (Cleveland: World Publishing, 1971): 431.
19 "Meeting of the Royal Astronomical Society," *Observatory* 53 (1930): 39.
20 G. Lemaitre, "Un Univers Homogène de Masse Constante et de Rayon Croissant, Rendant Couple de la Vitesse Radial des Nebuleuses Extra-Galactiques," *Ann. Soc. Sci. Brux.* 47 (1927): 49–56; translated in *Mon. Not. R. Astron. Soc.* 91 (1931): 483–490.
21 A. Friedmann. "Über die Krümmung des Raumes," *Z. Phys.* 10 (1922): 377–386.
22 A. Friedmann, "Über die Möglichkeit einer Welt mit Konstanter negativer Krümmung des Raumes," *Z. Phys.* 21 (1924): 326–332.
23 A. S. Eddington, "On the Relation between the Masses and the Luminosities of the Stars," *Mon. Not. R. Astron. Soc.* 84 (1924): 308–332.
24 J. Jeans, *Astronomy and Cosmology* (Cambridge: Cambridge University Press, 1922) Chapter 12.
25 H. Bondi and T. Gold, "The Steady-State Theory of the Expanding Universe," *Mon. Not. R. Astron. Soc.* 108 (1948): 252–270.

Chapter 4

1 E. Hubble, *The Realm of the Nebulae* (New Haven: Yale University Press, 1936): 120.
2 *Ibid.*, 123.
3 Private communication, E. Hubble to W. de Sitter, 23 September 1931 (Huntington Library).
4 Private communication, N. Mayall to E. Hubble, 16 March 1937 (Huntington Library).
5 Hubble, *Realm*, 122.
6 *Ibid.*

Problems
Section One

Chapter 1

1. Kant used Kepler's laws to determine the rate at which stars should be observed to move through the sky (proper motion). He showed that the proper motion could not be detected with the instruments of his era. Although this conclusion was correct, his reasoning was wrong. Why?
2. With a star atlas and by use of the method of star-gauging, can you find evidence that the Sun is eccentrically located in the Galaxy?
3. The quotation below is from the novel *The Confessions of Nat Turner* by William Styron, which won a Pulitzer Prize and is based on the events surrounding a slave uprising in Southampton County, Virginia, about 1831:

   "Well now in England there's a great astronomer name of Herschel. Know what an astronomer is? Yes? Well, there was a big write-up on him not long ago in the Richmond newspaper. What Professor Herschel has found is that this here star of our'n that we call the sun is but one of not thousands, not millions, but billions of stars all revolving around a big kind of cartwheel that he calls a galaxy. Fancy that Reverend . . ." he leaned forward towards me, and I could smell the apple-sweet perfume of his presence. "Fancy that! Millions and even billions of stars all floating around in the vastness of space, separated by distances the mind can't even conceive of. Why, Reverend, the light we see from some of these stars must of left there long before man hisself ever dwelt on earth! A million years before Jesus Christ!"

   Comment on the authenticity of the narrative. What size Galaxy did the narrator envision?
4. Discuss the lingering influence of Greek philosophy on astronomical theories of the 18th century.

Chapter 2

1. Since men (except for the astronauts) are confined to living on the Earth, we often forget that our perception of the universe has been strongly influenced by the place we live in. If we had evolved elsewhere— on a moon of Jupiter, for instance—our concept of the physical world might have been extremely different. Discuss how the following topics might have been approached by a Jovian civilization; geocentric model for the solar system; Joviocentric model for the solar system; the visual appearance of the Milky Way; calculations of stellar distances.
2. In time of war, the loyalty of a scientist is often divided between his state and his work. Astronomy especially has been affected by this division because international cooperation among observatories has been essential to some astronomical research programs. Amiable relations usually continue among scholars, astronomers included, almost irrespective of the ideological differences their governments may have. In war time, however, that cordiality and cooperation naturally diminish between scientists of combatant nations.

   Comment on the effect of hostilities on the personal attitudes of astronomers. Should a scientist's loyalty be first to his state or to his work and his colleagues? Are scientists, as a group, affected differently by war than are persons in other fields?

   What are the responsibilities of a scientist in times of war? In times of peace? Does a scientist have social responsibilities beyond those of the average citizen?

Chapter 3

1. From van Maanen's plate in the *Astrophysical Journal*, 44 (1916), p. 219, find the average velocity of the measured points, assuming M101 is comparable in size to (a) Shapley's diameter for our galaxy (100 kpc), (b) 1/10 Shapley's value.
2. Describe the difficulties that are inherent in making measurements like those attempted by van Maanen. What sources of error would you expect?
3. The following analysis of political revolution is from Thomas Kuhn's widely acclaimed *The Structure of Scientific Revolutions* (Chicago: Chicago University Press, 1962), p. 92:

   Political revolutions aim to change political institutions in ways that those institutions themselves prohibit. Their success therefore necessitates the partial relinquishment of one set of institutions in favor of another, and in the interim, society is not fully governed by institutions at all. Initially it is crisis alone that attenuates the role of political institutions. . . . In increasing numbers individuals become increasingly estranged from political life and behave more and more eccentrically within it. Then, as the crisis deepens, many of these individuals commit themselves to some concrete proposal for the reconstruction of society in a new institutional framework. At that point the society is divided into competing camps or parties, one seeking to defend the old institutional constellation, the other seeking to institute some new one.

Discuss the parallels between Kuhn's description of political revolution and the scientific revolution forced by Shapley's galactic theories. How realistic is the analogy? Where does it break down?

4. Some of Kuhns main points in *The Structure of Scientific Revolutions* are summarized in the following quotation:

The invention of other new theories regularly, and appropriately, evokes the same response from some of the specialists in whose area of special competence they impinge. For these men the new theory implies a change in the rules governing the prior practice of normal science. Inevitably, therefore, it reflects upon much scientific work they have already successfuly completed. That is why a new theory, however special its range of application, is seldom or never just an increment to what is already known. Its assimilation requires the reconstruction of prior theory and the reevaluation of prior fact, an intrinsically revolutionary process that is seldom completed by a single person and never over night.

Is this statement applicable to events surrounding the emergence of Shapley's galactic theory? Support your argument with examples.

Chapter 4

1. Shapley used the diameters of globular clusters to estimate their distances. What are the advantages of that method of determining distances in contrast to the schemes involving Cepheid variables and the 25 brightest stars? What are the disadvantages?

2. How important was technology to the development of astronomy from 1910 to 1920? What other factors had an effect? How?

3. In his last book, *Through Rugged Ways to the Stars* (New York: Scribners, 1969), Shapley reminisced about his professional life:

In a scientific way, I suppose my number one contribution was locating the center of our galaxy some 33,000 or more light-years from the sun; in other words, the "over-throw" of the helio-centric hypothesis of Copernicus.

Was Shapley's contribution comparable to that of Copernicus? What are the similarities, and what are the differences?

4. Shapley explained the asymmetric distribution of globular clusters by postulating that they outline the Galaxy and that we are located eccentrically. What other plausible explanations might be offered?

5. The dramatic strides made in galactic astronomy from 1800 to 1920 were achieved by astronomers in only a handful of countries—primarily England, Germany, Holland, Sweden, and the United States; indeed, from 1910 to 1930 many of the most significant developments in this field were made in only one nation—the United States. Were there logical reasons why these particular countries dominated this research? Since there is no evidence that American scientists are exceptionally brilliant, how did the United States become by 1930 the leader in galactic and extragalactic research?

6. Historians often assign precise dates to important discoveries—such as the "discovery" of America by Columbus. Could similar precision be applied to the astronomical discoveries made between 1910 and 1920? Give examples of specific scientific breakthroughs, for which exact dates can be given.

7. Compare the conflicts surrounding the "Great Debate" with those attending the discovery of quasars and pulsars. How would you predict that the quasar enigma will be resolved?

8. The development of astronomy, as discussed in this section, involved interactions among several people with divergent, strong personalities. From material in this text and the companion volume of Readings, describe the personalities you imagine the following scientists possessed: Curtis, Hale, Herschel, Kant, Maury, Leavitt, Shapley, van Maanen, and Wright.

Which of these persons do you believe you would have found the most likeable? The least likeable? How did their personalities affect the development of science?

To many people, the image of the scientist is that of a cool, brilliant, dispassionate man or woman who places his or her research above all else. How accurate is that stereotype? Compare the objectivity of scientists with that of other scholars, with that of the general public.

Problems
Section Two

Chapter 1

1. What advantages do photographs have over visual observations in astronomy?
2. Are there merits to empirical science? What are its benefits and dangers?
3. What problems are associated with determining stellar distances? Why is it necessary for astronomers to know those distances accurately?
4. "Astronomers have routinely ignored the data that their photoelectric measures provide on the terrestrial atmosphere. We have found that, in fact, these data are of value to the study of our atmosphere. . . ."

   This recent statement by Charleson, Hodge, Lucke, Mannery, and Snow indicates that standard astronomical techniques can be used to monitor the particulate pollution of our atmosphere, although earlier most astronomers were not aware of the possibility. In the light of this fact, should scientists be encouraged (or forced) to find or develop practical applications of their research techniques and results? How could they be encouraged? Should scientists' main concern be toward applied research?

Chapter 2

1. The results obtained by many scientists have vanished into obscurity because they were not widely circulated, as happened to those of Gyldén in 1871. Therefore, one might assume that scientists have a duty to each other to publish their results. At present, however, the amount of information to be disseminated has increased tremendously so that journals must restrict the number of articles accepted, thereby possibly forcing important information into obscurity. Comment on this dilemma. What possible solutions can you propose? If the present system of journal publication is to continue, what criteria must be used to decide whether to publish a particular article?
2. Why was the method of distance determination that utilized Oort's theory important to astronomers? Is there any other method with comparable features?
3. The development of early galactic-rotation theory depended on mutual aid among astronomers of several countries. Is it possible that increased cooperation between Soviet and United States scientists today might greatly advance astronomical research? If so, in what specific fields lie the best chances of improvement through international exchange of ideas?
4. Oort's theory and data indicated that the center of our Galaxy was in the same direction as Shapley had previously determined, yet the distances found by the two men differed considerably. Besides absorption, which had not been well determined by 1927, what explanation could you have provided for the differences between the theories, had you been an astronomer at that time? Can you think of observations to test your hypotheses?
5. Using Oort's constants ($A=15$ km/sec/kpc; $B=-10$ km/sec/kpc), determine if van Maanen's rotational measures are plausible (see Section Three).

Chapter 3

1. What physical problems arise when absorption of light by interstellar matter is evoked to solve Olbers' Paradox? How can these problems be overcome? (See Section Four.)
2. How did Slipher's observations of "stationary lines" differ from those of Hartmann? Why could Slipher draw far-reaching conclusions from his data, while Hartmann could not?
3. Which group of astronomers is more important to the development of astronomy—theoreticians or observers? Justify your answer with examples. Can the research of one group be relevant without the other group's results?
4. In a letter to J. C. Kapteyn, G. E. Hale mentioned that the "psychological moment" had arrived to consider the possibilities of general absorption in interstellar space. What astronomical developments in the few decades preceding 1914 could possibly have changed the receptiveness of astronomers to the concept of absorption?
5. Reasoning by analogy can be successful, as is evident in the case in which Curtis examined dark equatorial bands in spiral nebulae. Such a process of analysis can be dangerous, however. Identify a case in

which astronomers arrived at a false conclusion by utilizing analogies. Why did the analogy break down?

6. In 1909, Kapteyn stated that absorption would be the most natural explanation for the apparent "thinning out of stars for an increased distance from the sun." What caused him finally to ignore absorption? Were there psychological influences as well as physical arguments? How might the development of modern astronomy from 1920 to 1930 have changed if Kapteyn had continued in the belief he expressed in his statement of 1909?

7. Aside from navigation, does astronomy have any economic value? If so, explain that value, using examples. If not, why has astronomy been supported for thousands of years? Can astronomers, at present, justify requests for governmental financial support?

8. Herschel called dark nebulae "holes in the heavens." At what point in the history of astronomy could people have realized that the concept of such a "hole in the heavens" could not be valid? For what possible reasons did some astronomers still refer to "regions vacant of stars" in 1905?

Problems
Section Three

Chapter 1

1. Review the arguments of the protagonists in the "great debate" concerning the island-universe theory. Why did the determination of distances to a few spirals help settle the argument?

2. Consider the following two poems by William Bronk (from *Inside Outer Space: New Poems of the Space Age*, ed. R. Vas Dias [New York: Doubleday Anchor Books, 1970]: 50–52). Do you think the ideas expressed in them were prompted by the "great debate"?

Of the All With Which We Coexist

Looking around me, I see as far to one
sky as another. The limitations of the eye:
we know the sky goes farther. Yet instruments
give us the same view and absolve the eye.

If I am not central to the world, then it fails
to make any difference whatever I feel.
The universe is large: to be eccentric is to be
nothing. It is not worth speaking of.

If I am anything at all, I am
the instrument of the world's passion and not
the doer or the done to. It is to feel.
You, also, are such an instrument.

You speak of justice and injustice, and well you
    might.
You speak of grief, of ecstacy. This
is a cruel world and a gay one. *We* are. Feel.
There is nothing to do, to be done, to be done to.

The Various Sizes of the World

We all get used to the regular stars in time.
After the start of learning how far they are,
what distances from earth, and even more
what space they keep apart from star to star,
where centuries divide the closest star's faint
    light
from light beyond, the mind comes back at last
making the sky seem shallow like the earth
where, from the air, we see a city's lights
spread out across the surface crust below
in constellations we read without surprise.

The sky is a similar surface pierced with lights
until, another morning, the sensitive plate
of a telescope has fixed a light so far
we never knew, so huge that a galaxy needs
to hold it. What address ever really finds
us in the endless depths the world acquires?
The earth has mass to hold our own mass down,
and the huge sun holds earth as though
a whirled cord were taut with it. But the mind
responds to the pull of its own gravities.

The mind is shifted outward into space
beyond the sun, where the surface sky explodes
softly forever like an endless wind.
Out and back the mind, the slide of the rule.
Where shall we add the logarithm of what
to find the actual product of any hour?
What point can fix the decimal of space
that joins the least remoteness of the earth
by tiny increments to the last star?
No, here's an incongruous world, too large, too
    far.

3. Would Slipher's drift theory imply that spirals are island universes? Explain.

Chapter 2

1. Why is it that in themselves neither angular diameters nor apparent magnitudes of galaxies give reliable distance measures? Why is this fact important?
2. Why would Slipher's large radial velocities for spirals imply that measurable proper motions for the spirals as a whole should be found? Refer particularly to Shapley's method of calibrating the period-luminosity relation.
3. Discuss the actual evidence before 1924 (other than van Maanen's results) pertaining to spirals as island universes. Could any of this evidence definitely have resolved the question? Could van Maanen's results definitely have resolved the question?

Chapter 3

1. By the early 1920s, two types of observations concerning motions in spirals were available—Slipher's spectroscopic velocities and van Maanen's proper motions. Were the implications of these two sets of findings complementary or contradictory with respect to rotations? Explain.
2. No other astronomer could verify van Maanen's findings of internal motions in spirals. Comment on whether this fact reflects on the reliability of van Maanen's work. What was the opinion of astronomers at the time?
3. Until 1925, no theory supported the view that spirals were island universes. Demonstrate this by briefly summarizing the theories of Laplace, Chamberlain and Moulton, and Jeans.

Chapter 4

1. Comment on the first two of the three assumptions underlying Hubble's studying of Cepheids in spirals. Are the assumptions reasonable? Are they correct? If not, what corrections should be made?
2. Do you agree with Sandage's statement about how Hubble's discovery settled the "great debate"? Why did the argument continue after 1924? When did it finally end?
3. Should astronomers have expected to find Cepheids in spirals? Why?
4. Hubble began his famous book, *The Realm of the Nebulae*, with the following paragraph:
Science is the one human activity that is truly progressive. The body of positive knowledge is transmitted from generation to generation, and each contributes to the growing structure. Newton said, "If I have seen farther it was by standing on the shoulders of giants." Today [1936], the least of the men of science commands a wider prospect. Even the giants are dwarfed by the great edifice in which their achievements are incorporated. What a Newton might see today, we do not know. And tomorrow, or a thousand years hence, even our dreams may be forgotten.
Comment on the validity of this statement in relation to Hubble's own work.

Chapter 5

1. In order to detect proper motions from photographic plates, why is it necessary that the plates be separated by a long interval of time? Would a longer interval make the observations easier?
2. Before Hubble's discovery of Cepheids in 1924, Jeans had agreed with van Maanen's work. In Chapter 2, we saw that Jeans also had agreed with Hubble's work at the same time; yet after 1924, Hubble and van Maanen adamantly disagreed.

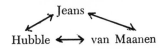

Clearly describe these influences and determine if there are any contradictions.
3. Consider Shapley's comments about Hubble and van Maanen:
van Maanen at once became a friend of Mr. Hale. Van Maanen was aggressive and he was sociable. He could go to dinner and soon have the whole table laughing. He was a social success. People liked him. . . .
Hubble disliked van Maanen from the time he himself arrived on Mount Wilson; he scorned him. Hubble just didn't like people. He didn't associate with them, didn't care to work with them.
Discuss the influence of personalities and social abilities on scientific research.
4. It has recently been shown (J. O. Stenflo, "Hale's Attempts to Determine the Sun's General Magnetic Field," *Sol. Phys.* 14 [1970]: 263–272) that van Maanen's study of the solar magnetic field was also subject to systematic errors, thus making it invalid. Does this discovery reinforce the conclusion in this section concerning the nature of van Maanen's error in determining internal motion in spirals? Explain.

Chapter 6

1. a) Using the Herschel system, classify each of Wolf's drawings.
   b) Using Hubble's system, classify each of Wolf's drawings.

c) Compare your results with those of other students in your class. Do you get greater agreement by using Herschel's scheme or Hubble's? Why?
2. Why is classification important?
3. Because of general anti-German attitudes following World War I, the I.A.U. originally did not have any German members. Did this have any effect on the avowed aims of the organization? Explain.

4. How obvious is the distinction between non-galactic nebulae and galactic nebulae with regard to
   a) position in the sky?
   b) spectrum?
   c) velocity?
   d) appearance?
5. What evidence today indicates that galaxies do *not* evolve from one type to another, as Hubble thought?

Problems
Section Four

Chapter 1

1. Lord Kelvin, a great physicist of the nineteenth century, said, "I cannot satisfy myself until I can make a mechanical model of a thing. If I can make a mechanical model, I can understand it. As long as I cannot make a mechanical model all the way through, I cannot understand." Contrast this viewpoint with that of the twentieth century.
2. Give an argument based on common experience that could prove the existence or nonexistence of an absolute reference frame such as the aether.

Chapter 2

1. Relativity allows a man to define his position as the center of the universe; i.e., each reference frame is as good as another so that any one chosen is valid. Comment on the differences between this egocentric discussion and the dogmatic pre-Copernican theories.
2. Einstein stated, "By an application of the theory of relativity to the taste of the reader, today in Germany, I am called a German man of science and in England, I am represented as a Swiss Jew. If I come to be regarded as a bête noire the description will be reversed." Science has rarely been completely free of politics. Should it be?
3. "It's not science that must be restricted, but rather the scientific investigators and teachers;

only scientifically talented men who have pledged their entire personalities to the nation, to the racial conception of the world, and to the German mission will teach and carry on research at the German Universities."

This sentiment, expressed during the Nazi reign in Germany, would find little support today. On the other hand, is the movement to restrict science to the mission of social and environmental improvement guilty of the same intellectual supression? Can basic research be neglected in favor of applied science without harmful effects?
4. Find the point on Earth from which you can travel one mile south, one mile east, and then one mile north, and arrive back at your starting point. From what points on Earth could you follow the same directions and finish one mile west of your starting point?
5. Symmetry is an important factor in the development of relativity. What reason exists for the elevation of symmetry to a basic scientific precept?
6. Much of relativity was developed intellectually without the impetus of experimentation. Should the theory then be considered a philosophical or a scientific achievement?
7. Does relativity invalidate Newton's Laws? How do you think Einstein would have answered this question?
8. "This is indeed a mystery," remarked Watson. "What do you imagine it means?"
   "I have no data yet. It is a capital mistake

to theorize before one has data. Insensibly, one begins to twist facts to suit theories, instead of theories to suit facts."

—*Sherlock Holmes*, by A.C. Doyle

Einstein apparently violated Sherlock Holmes' dictum; yet, later experiments seem to have confirmed his theory. What are the advantages of Einstein's approach? Is his method scientifically valid? Can one trust experimental results determined after the presentation of a theory?

9. Some people believe that cosmology is an exercise in futility—no scientific confirmation is possible. Is that belief valid today? Was it valid before 1930?

10. The works of Lobochevsky and Bolzan of Russia were unknown to West Europeans, mainly because of the language barrier. Today, thanks to organized translation services, language is usually not a difficult barrier to overcome. Other barriers, however, hinder the dissemination of information. What are they, and how can they be eliminated?

11. The geometric interpretation of gravity or of relativity increases the mathematical complexity. Why have people been willing to accept the difficulties involved in this formalism?

12. Einstein's name is usually associated with the formulation of the theory of relativity, yet he depended strongly on the non-Euclidean geometries developed previously. Should the geometers share equal credit?

13. What possible reasons could have influenced Einstein to support a static-universe theory of cosmology?

Chapter 3

1. Although most of the radial velocities found by Slipher indicated motions away from the Milky Way, some (most notably M31) showed motion *toward* the Milky Way. How could this be explained in a way consistent with the concept of an expanding universe?

2. In December 1919, the Royal Astronomical Society decided to award its Gold Medal to Einstein. A few days later, however, the Society officials changed their minds, and for the first time in 30 years no Gold Medal award was made. Einstein finally did receive the award in 1926. Could this remarkable event have occurred because of the political climate of the times?

3. Because of Silberstein's spurious results, Hubble had to be cautious in announcing the velocity-distance relation. But was his distinction between observation and theory as clear as in the case of classification of galaxies? Could Hubble's tremendous reputation at the time of the velocity-distance discovery have made a difference in the acceptance it received?

4. Can the introduction of the cosmological constant be considered an ad hoc assumption?

5. Why was the velocity-distance relation *alone* not enough to cause astronomers to discard static models of the universe?

6. Does the velocity-distance relation require a definite beginning of the universe?

7. Both the steady-state and the big-bang theories require an ad hoc assumption about creation. Describe the differences in the assumptions made by the two models. Why does the assumption in the steady-state model imply that energy is not conserved, whereas the assumption in the big-bang model allows energy to be conserved?

8. If it were discovered that the redshifts do not represent velocities, could the steady-state theory require continuous creation? What would then be the differences between the big-bang and steady-state models?

9. Could Einstein's or de Sitter's model be related to a steady-state theory?

10. The velocity-distance relation can be stated as v=Hd, where v=velocity, d=distance, H = Hubble's constant—50 km/sec/Mpc. Compare this with the simple equation of

motion $V = \dfrac{d}{t}$, where v and d are as above

and t=time. Does the Hubble constant tell us anything about the age of the universe? Explain. Determine an age for the universe from the above value of H. How does this compare with the age of the Earth?

11. Hubble originally assigned a value of 500 km/sec/Mpc to the constant H. Today, an acceptable value is 50 km/sec/Mpc. What significance does the reduction in value have in relation to the age of the universe? What discoveries caused the value to be reduced?

12. Several times in the history of astronomy, two conflicting views have come into dramatic opposition. The resulting controversies were solved only by new advances in observations. Three of the most significant of these cases involved the Copernican theory, the island-universe theory, and the big-bang theory. Describe the opposing views and the chief spokesman for each view. For the first two cases,

describe the observational advance that settled the dispute. What is the current status of the third case? Does it parallel the development of the first two?

Chapter 4

1. There have been strong arguments against interpreting the redshifts of galaxies as radical velocities, particularly in the case of quasars. Comment on the evidence in astronomy today concerning this argument.

2. Suppose it were discovered that light does lose energy merely by traveling through space. What effect would this have on the concept of conservation of energy?

3. One consequence of the general theory of relativity is that light emitted from a massive object should be redshifted, owing to the gravitational attraction of photons by the masses. Could this principle lead to a redshift-distance relation?

Bibliography

I. Biographical Information

1. J. Berstein, *Einstein* (New York: Viking Press, 1973).
2. R.W. Clark, *Einstein: The Life and Times* (Cleveland: World Publishing, 1971).
3. C.C. Gillispie, ed. *Dictionary of Scientific Biography* (New York: Charles Scribner's Sons, 1970ff).
4. B. Hoffmann and H. Dukas, *Albert Einstein: Creator and Rebel* (New York: Viking Press, 1972).
5. H. Shapley, *Through Rugged Ways to the Stars* (New York: Charles Scribner's Sons, 1969).
6. H. Wright, *Explorer of the Universe: A Biography of George Ellery Hale* (New York: Dutton & Co., 1966).
7. H. Wright, J. Wurnaw, and C. Weiner, *The Legacy of George Ellery Hale* (Cambridge: MIT Press, 1972).

II. Major Works on the History of Astronomy

1. A. Berry, *A Short History of Astronomy* (reprinted by New York: Dover Publications, Inc., 1966). Original publication (London: J. Murray, 1898).
2. A. Clarke, *A Popular History of Astronomy* (London: Adams and Charles Black, 1893).
3. M. Hoskin, *William Herschel and the Construction of the Heavens* (New York: W.W. Norton, 1964).
4. S.L. Jaki, *The Paradox of Olbers' Paradox* (New York: Herder and Herder, 1969).
5. S.L. Jaki, *The Milky Way* (New York: Neale Watson, 1972).
6. H. MacPherson, *Modern Cosmologies: A Historical Sketch of Researches and Theories Concerning the Structure of the Universe* (Oxford, Eng.: Oxford University Press, 1929).
7. M.K. Munitz, ed. *Theories of the Universe* (Glencoe, Ill.: The Free Press, 1957).
8. J.D. North, *The Measure of the Universe: A History of Modern Cosmology* (Oxford, Eng.: Oxford University Press, 1965).
9. O. Struve and V. Zebergs, *Astronomy of the 20th Century* (New York: Macmillan Co., 1962).
10. C. Whitney, *The Discovery of Our Galaxy* (New York: Alfred A. Knopf, 1972).

III. Works on the History of Astronomical Observatories

1. B.Z. Jones and L.G. Boyd, *The Harvard College Observatory* (Cambridge: Harvard University Press, 1971).
2. *Lick Observatory* (Berkeley: University of California Press, 1961).
3. *Lowell Observatory* (Flagstaff, Ariz. Chamber of Commerce).
4. A.E. Whitford, "Astronomy and Astronomers at the Mountain Observatories," *Proc. Int. Conf. Ed. Hist. Astron.* (published in the *Annals of the New York Academy of Sciences,* 1972).
5. D.O. Woodbury, *The Glass Giant of Palomar* (New York: Dodd, Mead & Co., 1939).

IV. Important Published Research Articles from the Period

A. Galactic Models

1. H.D. Curtis and H. Shapley, "The Scale of the Universe," *Bull. Nat. Acad. Sci. 2* (1921): 171–217.
2. I. Kant, *Universal Natural History and Theory of the Heavens* (Ann Arbor, Mich.: University of Michigan Press, 1969).
3. J.C. Kapteyn, "First Attempt at a Theory of the Arrangement and Motion of the Sidereal System," *Astrophys. J. 55* (1922): 65–91.
4. H. Shapley, "Studies Based on the Colors and Magnitudes in Stellar Clusters," *Astrophys. J. 45* (1917): 118–141 and 164–181; *46* (1917): 64–75; *48* (1918): 89–124, 154–181, and 279–294; *49* (1919): 24–41, 96–107, 249–265, and 311–336; *50* (1919): 42–49 and 107–140.
5. H. Shapley, *Star Clusters* (New York, McGraw-Hill, 1930).

B. Absorption

1. E.E. Barnard, "On the Vacant Regions of the Milky Way," *Popular Astronomy 14* (1906): 579–583.
2. E.E. Barnard, "Some of the Dark Markings on the Sky and What They Suggest," *Astrophys. J. 43* (1916): 1–8.
3. H.D. Curtis, "Absorption Effects in the Spiral Nebulae," *Publ. Astron. Soc. Pac. 29* (1917): 145–146.
4. J.C. Kapteyn, "On the Absorption of Light in Space," *Astrophys. J. 30* (1909): 284–317.
5. R. Trumpler, "Preliminary Results on the Distances, Dimensions, and Space Distribution of Open Star Clusters," *Lick Obs. Bull. 14* (1930): 154–188.

C. Nature of Spiral Galaxies

1. H.D. Curtis, "Novae in Spiral Nebulae and the Island Universe Theory," *Publ. Astron. Soc. Pac. 29* (1917): 206–207.
2. E.P. Hubble, "Extra-Galactic Nebulae," *Astrophys. J. 64* (1926): 321–369.
3. E.P. Hubble, "Cepheids in Spiral Nebulae," *Observatory 48* (1925): 139–142.
4. E.P. Hubble, "Angular Rotations of Spiral Nebulae," *Astrophys. J. 81* (1935): 335–336.

5. E.P. Hubble, *The Realm of the Nebulae* (New Haven: Yale University Press, 1936).
6. K. Lundmark, "Internal Motions of Messier 33," *Astrophys. J. 63* (1926): 67.
7. A. van Maanen, "Preliminary Evidence of Internal Motion in the Spiral Nebula Messier 101," *Astrophys. J. 44* (1916): 210–228.
8. A. van Maanen, "Investigations on Proper Motion, Tenth Paper: Internal Motions in the Spiral Nebula Messier, 33, NGC 598," *Astrophys. J. 57* (1923): 264–278.
9. A. van Maanen, "Angular Rotations in Spiral Nebulae," *Astrophys. J. 81* (1935): 336–337.
10. H. Shapley, "More on the Magnitudes of Novae in Spiral Nebulae," *Publ. Astron. Soc. Pac. 29* (1917): 213–217.
11. H. Shapley and H.D. Curtis, "The Scale of the Universe," *Bull. Nat. Res. Coun. 2* (1921): 171–217.
12. V.M. Slipher, "The Radial Velocity of the Andromeda Nebula," *Lowell Obs. Bull. 2* (1913): 56–57.
13. V.M. Slipher, "The Direction of Nebular Rotations," *Lowell Obs. Bull. 2* (1914): 66.

D. *Relativity*

1. A. Einstein, H.A. Lorentz, H. Minkowski, and H. Weyl, *The Principle of Relativity*, translated by W. Perrett and G.B. Jeffrey. (New York: Dover Publications, 1923).
2. G. Holton, "Einstein, Michelson, and the Crucial Experiment," *Isis 60* (1969): 133–197.
3. B. Russell, *The ABC of Relativity* (New York: New American Library, 1959).
4. J.H. Smith, *Introduction to Special Relativity* (New York: W.A. Benjamin, Inc., 1965).
5. L.S. Swenson, "The Michelson-Morley-Millar Experiments before and after 1905," *J. Hist. Astron. 1* (1970): 56–78.
6. H. Weyl, *Space Time Matter* (New York: Dover Publications, 1950).

E. *The Velocity-Distance Relation and Cosmology*

1. A.S. Eddington, *The Expanding Universe* (Cambridge, Eng.: Cambridge University Press, 1933).
2. E.P. Hubble, "A Relation Between Distance and Radial Velocity Among Extra-Galactic Nebulae," *Proc. Nat. Acad. Sci. 15* (1929): 168–173.
3. E.P. Hubble, *The Observational Approach to Cosmology* (Oxford, Eng.: Oxford University Press, 1937).

V. Other Bibliographies

1. *Astronomischer Jahresbericht* (issued annually from 1889 to 1968).
2. *Astronomy and Astrophysics Abstracts* (1969-present).
3. *Isis Critical Bibliography* (issued annually since 1913).

Index